Skills in Advanced Biology

Volume 2

Observing, Recording and Interpreting

This book is due for return on or before the last date shown below.

Don Gresswell Ltd., London, N.21 Cat. No. 1208

DG 02242/71

Skills in Advanced Biology

Volume 2

Observing, Recording and Interpreting

J W Garvin BSc DipEd DMS CBiol MIBiol

Head of Science, Cambridge House Grammar School for Girls, Ballymena

J D Boyd BSc CertEd CBiol MIBiol

Head of Biology, Cambridge House Grammar School for Boys, Ballymena

Stanley Thornes (Publishers) Ltd

First published in 1990 by:
Stanley Thornes (Publishers) Ltd
Old Station Drive
Leckhampton
CHELTENHAM   GL53 0DN
England

British Library Cataloguing in Publication Data

Garvin, J. W.
Skills in advanced biology.
Vol. 2, Observing, Recording and Interpreting.
1. Biology. Laboratory techniques
I. Title   II. Boyd, J. D.
574'.028

ISBN 0-85950-817-X

Typeset by Tech-Set, Gateshead, Tyne & Wear
Printed and bound in Great Britain at The Bath Press, Avon

To our respective XX partners

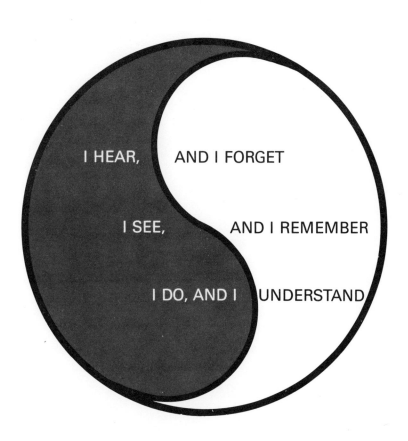

I HEAR,   AND I FORGET

I SEE,   AND I REMEMBER

I DO, AND I   UNDERSTAND

Contents

SKILLS AND CONTENTS MATRIX

It is difficult to decide on the best way to arrange the materials. We have designed the volume so that you can go across the grid shown below, or down the grid, depending on how the materials are incorporated into a course.

	1.1 OBSERVING	1.2 RECORDING	1.3 INTERPRETING
1 INTRODUCING THE SKILLS	1.1.1 Seeing and perceiving 1.1.2 Observing and knowledge 1.1.3 Observing and hypotheses 1.1.4 Observing and investigating 1.1.5 Extrasensory receptors 1.1.6 Microscopy 1.1.7 Remote sensing 1.1.8 3-D from 2-D 1.1.9 Magnification and scales	1.2.1 Drawings and diagrams 1.2.2 Tables 1.2.3 Graphs	1.3.1 Drawings and diagrams 1.3.2 Graphs 1.3.3 Statistics
	1 PRINCIPLES	2 LEARNING THE SKILL	3 APPLYING THE SKILL
2 DRAWING	2.1.1 Biological drawings 2.1.2 Labelling 2.1.3 Annotating	2.2.1 The cell 2.2.2 Angiosperm histology 2.2.3 Mammalian histology 2.2.4 Lower animals	2.3.1 The cell – TEMs 2.3.2 Protists 2.3.3 Angiosperm structures 2.3.4 Hydrophytes and xerophytes 2.3.5 Primroses 2.3.6 Parasites 2.3.7 Mammal organs
3 COUNTING	3.1.1 Aids to counting 3.1.2 Sampling 3.1.3 The haemocytometer	3.2.1 Counting sheep 3.2.2 Eggs in the trematode *Microphallus pygmaeus* 3.2.3 *Drosophila:* male and female 3.2.4 Haemocytometry	3.3.1 Sampling weeds in a field 3.3.2 Maize cobs 3.3.3 Population growth in yeast 3.3.4 *Sordaria* asci 3.3.5 Human chromosomes: abnormalities 3.3.6 *Drosophila* crosses 3.3.7 Mitochondria in trematode cercaria tail
4 MEASURING	4.1.1 The unaided eye 4.1.2 Micrometry 4.1.3 Photomicrography 4.1.4 Stereology	4.2.1 Micrometry and graticules 4.2.2 Stereological techniques 4.2.3 Chromatography	4.3.1 Lengths of virus particles 4.3.2 Banding in collagen fibres 4.3.3 Sizes of cells 4.3.4 Bean root cells 4.3.5 *Chlorella:* areas and volumes 4.3.6 Plasmolysis in onion epidermal cells 4.3.7 Capillaries in the lung 4.3.8 Heart muscle 4.3.9 The tongue 4.3.10 Other plates
5 OBSERVATIONAL ANALYSIS	5.1.1 Thin-layer chromatography 5.1.2 Electrophoresis 5.1.3 Autoradiography 5.1.4 DNA sequencing	5.2 Classification and keys	

Preface

To the Teacher

As a result of the introduction of GCSE, many changes are taking place in GCE Advanced-level syllabuses and their equivalents. They now list examination objectives which require candidates to possess a variety of skills, such as observing, recording, interpreting, analysing, investigating, etc. In order for students to become competent biologists and to do themselves justice in examinations they must develop these skills by careful learning and practice.

As well as these changes in the examinations, more students are staying on at school or are attending college to study A-levels, with the consequence that the ability range is widening and the variety of subjects studied with biology is increasing.

To overcome these problems a series of skill-based active learning volumes has been designed. These volumes are not actually textbooks or even courses; they consist of materials that can be used as flexible resources in whatever way that you want to use them. They do not require any prior knowledge except for the earlier volume of the series. They consist mainly of procedures whereby the skills are learned using examples, and then applied using exercises. These volumes lend themselves to individual study or as an integral part of an A-level course or in combination; they can be used as an aid to teaching and as a back up, and they often include testing of the concepts. The opportunity is available for students to work at their own pace and in their own time when necessary. The range of topics has been made deliberately wide so that the examples and exercises show how the techniques can be applied in many different situations.

Volume 1, titled *Dealing with Data* (from Stanley Thornes (Publishers) Ltd) is mainly concerned with the mathematical aspects of biology. Section 1 describes the types of data; Section 2, organising and presenting data (tables and graphs); and Section 3, analysing data (graphs and statistics). These skills are necessary to answer satisfactorily some of the questions in Volume 2.

Volume 2 Observing, Recording and Interpreting

This second volume in the series is an introductory course in the application of the techniques of drawing, counting and measuring. These skills are not normally taught in any structured or progressive way yet the new examinations assess such skills.

The **Student's Text** introduces each technique and teaches the student to apply them in a variety of biological situations by using examples and exercises. Biological experiments and investigations often produce data which are unreliable. This is not only because biological material is variable, but also because the student needs to repeat a procedure many times before being able to produce consistent and valid data. It is virtually impossible for a student to be given a recipe and produce a good cake at the first attempt. Practice in the required techniques is essential – certain skills have to be developed and then combined in order to achieve consistency in data generation. This book mimics practical situations, many of which are commonly found in biology courses. You can use these so that your students can practise the relevant skills before carrying out an investigation; your students will then have a better idea of what they may see, what they have to do and how to do it. The materials are also suitable for use after a practical has been carried out if the data is inconsistent, so that everyone can work on the same materials. Opportunities also exist to highlight deficiencies in technique or sources of experimental or observational error. The skills, once developed, can be used in a multitude of new situations and your students will be much more confident in their tackling of them. The usual photographs in most books are either labelled or accompanied by labelled diagrams. Our photographs are unlabelled so that the structures have to be identified using the information given in the text or by reference to other texts. Finally, many of the exercises can be used for homework or revision, and are invaluable to anyone engaged in project work; the text is also a rich source of ideas for investigations.

The *Teacher's Supplement* is an essential complement to the Student's Text. It contains answers to most of the questions and also acts as a guide where difficulties might arise. It contains master sheets of plates that need to be photocopied on to acetate sheets. Photocopying from the Supplement is permissible within the purchasing institution.

To the Student

The day is past when you could sit down at a book, cram in as much as possible in as short a time as possible, and hope to succeed in the examination. It is still important that you have a good background knowledge of the subject, but the trend is towards questions that test a whole range of skills – you need to be able to observe, record, interpret, analyse, investigate, etc. Initially, you probably feel very insecure with such questions because you have not developed the appropriate skills, but once you have acquired them you will feel much more confident. The skills have to be learned with understanding and then practised repeatedly. The skills will not only be essential for success in your examinations, they will also be very useful for further studies in biology and other subjects.

This second volume in the series is concerned with questions that test your ability to observe, record and interpret. Observing in biology involves drawing, counting and measuring, and you are taught to do these in the most accurate and efficient ways possible. Each skill is introduced, and then you are taught them by the use of a wide range of examples and exercises. Finally, you are shown how to apply these skills in a variety of situations. You should then be able to apply the appropriate skill in any new situation that arises. You can, if necessary, work on your own, in your own time and at your own speed. If you have any difficulties, consult your teacher.

We hope that you will find the work interesting and stimulating.

All the best

JWG, JDB   1990

Acknowledgements

Other people inevitably become involved when one attempts to translate ideas into book form. We have been very encouraged to find so many that were willing to assist us when necessary. We would particularly like to thank:

The following members of staff from Queen's University, Belfast: Professor B. Bridges and Mr K. Lee, Department of Physiology, Medical Biology Centre, for supplying us with many light and electron micrographs; Dr Philip Earle, Northern Ireland Centre for Genetic Engineering, Medical Biology Centre, for help with the section on DNA sequencing; Dr A. Ferguson, Department of Zoology, for electrophoretic gels; Dr T. W. Frazer, Microscopy Unit, Agriculture and Food Science Centre, for his continued interest and supply of electron micrographs; Mr G. McCartney, EM Unit, Department of Botany, for his wide range of suitable electron micrographs; Dr I. Hickey, Department of Genetics, for photographs of chromosomes.

Our lecturers at Queen's University of many years ago, Dr B. Gunning and Dr M. Steer, now respectively at the National University of Australia and University College Dublin, who taught us the techniques of stereology.

Dr S. W. B. Irwin, Department of Biology and Biomedical Sciences, University of Ulster at Jordanstown, for his interest and help in obtaining suitable materials, in particular electron micrographs.

Dr R. Storey, Biology Department, School of Science, Humberside College of Higher Education, for his interest and assistance with the section on remote sensing.

BKS Surveys Ltd for supplying aerial photographs.
Mr T. J. and Mr K. Gawn, for allowing their sheep to be photographed.
Mr Jim Kelso and Mr James Garvin for photographic assistance.
Mitchell Sneddon, for his pencil drawings.
Our colleague, J. A. McMillan for his useful comments.
To all our past students.
To our respective wives, Betty and Ruth, for putting up with us, just!

Apologies to anyone that we have left out; it was entirely unintentional. Also, any mistakes in this book are our responsibility.

Finally we would like to express our appreciation to those at Stanley Thornes who put it all together.

JWG, JDB   1990

The authors and publishers are grateful to the following for permission to reproduce material:

Edward Arnold: Figure 2.1(a) from K. G. Brocklehurst and B. J. Davies, *Photobiological Studies No. 2: Plant Histology* (Hodder & Stoughton, 1978)
Biofotos, Heather Angel: Figures 2.22, 2.58; Plates 2.17, 4.6, 5.2(d), (e), (f), (g),
Biophoto Associates: Figures 2.18(a), 2.24, 2.25(a), (b); 2.26(a), (b), 2.27(a), (b), 2.28, 2.29(a), (b), 2.30(a), (b), (c), 2.41, 2.42, 2.43, 2.44, 2.45(a), (b), 2.47, 2.55; Plates 2.18, 2.19, 2.20, 2.21, 2.22, 2.23(a), (b), 2.24(a), (b), 2.25(a), (b), 2.26(a), 2.28(a), (b)
BKS Surveys Ltd: Figure 1.33; Plates 1.1(d) and 3.5

Professor B. Bridges and K. Lee: Figures 2.2(a), 2.13(a), 2.39, 2.40, 2.49, 2.50, 2.51; Plates 2.3(a), (b), 2.4(a), (b), 2.8, 2.9, 2.11, 2.12, 2.29(a), (b), (c), (d), (e), 2.30(d), (e), (f), 2.31(a), (b), (c), (d), 2.32(a), (b), (c), (d), (e), (f), (g), (h), (i), 4.3, 4.9, 4.10, 4.11, 4.12, 4.13

Bruce Coleman Ltd: Figure 1.1 (Norman Tomalin); Plates 2.5(a) and 2.6(a) (Kim Taylor), 2.14, 2.15 and 2.16 (Frieder Sauer), 5.2(a) (Jane Burton), 5.2(b) (M.P.L. Fogden), 5.2(c) (A. Pasieka)

Department of Trade and Industry: Figure 1.35 (redrawn)

Dr P. Earle: Plates 5.1(a), (b)

Dr A. Ferguson: Figures 5.2, 5.3

Dr T.W. Frazer: Figure 1.27(b); Plate 4.2

Dr I. Hickey: Figures 3.14, 3.16

Eric Hosking: Plates 2.7(a) (G. E. Hyde), 5.2(h)

Dr S. W. B. Irwin: Figures 2.12(a), 3.3; Plates 1.3(b), 2.30(a), (b), (c), 3.13(a), (b)

Dr S. W. B. Irwin and Dr G. McKerr: Plates 2.26(b), 2.27

Dr S. W. B. Irwin and J. G. Rea: Figure 3.19; Plates 3.14 (a), (b)

Jim Kelso: Figures 1.11, 1.13, 1.14, 2.6, 4.1; Plates 1.2(c), 3.6(a), (b), 3.7(a), (b)

Mr G. McCartney: Figures 1.45, 2.17(a), 2.20(a); Plates 1.1(a), (b), (c), 1.3(a), 1.4(a), (b), (c), 2.10, 4.4

Dr G. McKerr: Plate 1.3(d)

Oxford Scientific Films: Figures 1.20, 1.37, 2.23, 2.56, 2.57; Plates 1.3(c), 1.4(d), 2.6(b), 2.7(b)

Mitchell Sneddon: Figures 2.1(b), 2.2(b)

Dr M. Steer and Dr B. Gunning: Figures 4.3, 4.4, 4.7, 4.8, 4.9, 4.10, 4.11, 4.12 (all redrawn)

Dr R. Storey: Figures 1.30, 1.31, 1.32, 1.36, 1.38, 1.39, 1.40(a), (b) (all redrawn except for 1.40(a))

University College of North Wales, Bangor, The School of Plant Biology: Figure 2.11(a)

Weidenfeld and Nicholson for allowing us to redraw Figure 1.2 from R. L. Gregory and E. M. Gombrich (Eds.), *Illusion in Nature and Art* (1970)

Every effort has been made to contact copyright holders and we apologise if any have been overlooked.

1 Introducing the Skills

The traditional view of scientific method, first proposed by Francis Bacon (1561–1626), claims that science begins with observations. The main basis of Bacon's ideas is that the truth lies all around us, and if we observe nature, then the truth will be revealed to us. Observation was thus considered to be a secure base from which all knowledge is derived and so knowledge in its turn became centrally important.

This attitude tended to be reinforced by textbooks and courses where knowledge for its own sake was considered to be of the utmost importance. Examinations tended to add fuel to this situation by placing too much importance on the recall of knowledge. Fortunately this situation has changed so that now syllabuses list skills and abilities that will be tested by the examinations. They also assess if candidates have the ability to apply the knowledge they do possess to new situations and problems. This attitude reflects much more accurately how scientists themselves operate and it is a welcome change.

The truth is not all around us just waiting to let us know all about it. In biology, we can never know in advance which observations are important and relevant and which are not. How then do scientists discover the truth, or at least a nearer approximation to the truth? Observations are important, since they are accepted as facts and are thus given high status. However, the day to day work of scientists consists not of hunting for facts but of testing hypotheses. These hypotheses are hunches, or tentative explanations, ideas of what the truth might be, and they arise creatively in the human mind.

1.1.1 Seeing and perceiving

Figure 1.1
The human eye

When a scientist observes something, it is only a small part of a large number of possible observations that could have been made; the observations made depend on who is doing the observing. We interpret observations in terms of what knowledge we possess and what our previous experiences have been; observation involves perception. Sir Alexander Fleming's discovery of penicillin is often described as accidental. A very large number of people could have seen whatever it was that he saw without perceiving it as he did. It was his perception of what he observed that led to the discovery because he was prepared for it; he had the specialist knowledge.

The optical model of the mind
Many of the words used to describe mental function are connected with the sense of sight. Perception refers both to seeing and to intuition. Insight, foresight and vision all imply intelligence.

1.1.2 Observing and knowledge

Images on the retina are converted into nerve impulses which travel via the optic nerve to the visual cortex of the brain, where thc object is 'seen'. These nerve impulses are the result of neural processing which has taken place in the retina itself: one optic nerve fibre may be represented by over 125 photo-receptors. It is not the eyes that see, nor for that matter the ears that hear, but the brain that sees and hears. What you 'see' is an interpretation of an object by the braın which is coloured (sic) by your perception of reality which has been built up in your brain since you were born, or even before. Is this actually what happens?

If you have read the previous paragraph without noticing any mistakes then you have a good illustration of what is meant. Actually there were two deliberate mistakes – the first one is on the second line where a 'c' replaces an 'e' in the word 'the'; the second error is on line seven, where the dot over the 'i' in the word 'brain' is missing.

The image of the world on the retina is upside-down and right–left reversed yet we see the world the right way up and the right way around, because our experience tells us that this is how it is. Every observation depends on what you already know.

Figure 1.2
What do you see?

To illustrate the principle of prior knowledge having an effect on what we see, look at the drawing in Figure 1.2 which was taken from a photograph having no grey tones. Do you see anything more than a mass of unintelligible blotches? You have no knowledge that would help you with your perception. However once you know exactly what to look for, you should be able to see it clearly, and you will generally see it whenever you look at the drawing again.

Examine Plate 1.1 which gives four photographs of different objects (a), (b), (c) and (d).

1 Write down what you think they are.

Since you have no prior knowledge relating to these images, interpretation can be difficult. A few clues might help, for example photographs (a) and (b) show quite common objects and they were taken with a scanning electron

PLATE 1.1

a

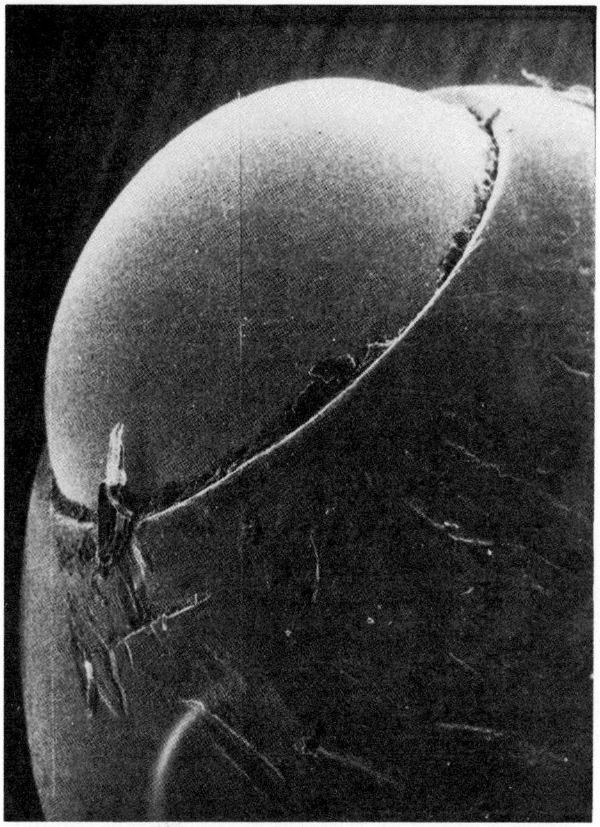

b

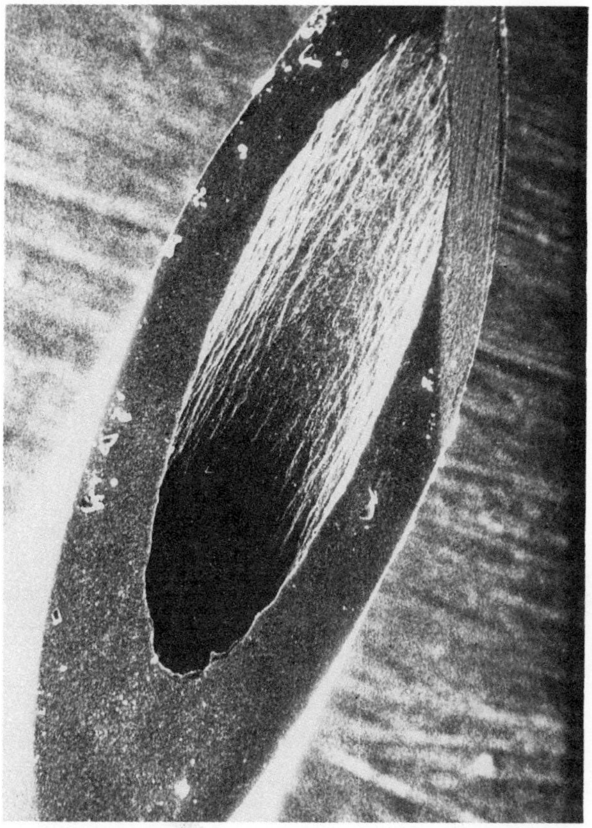

c

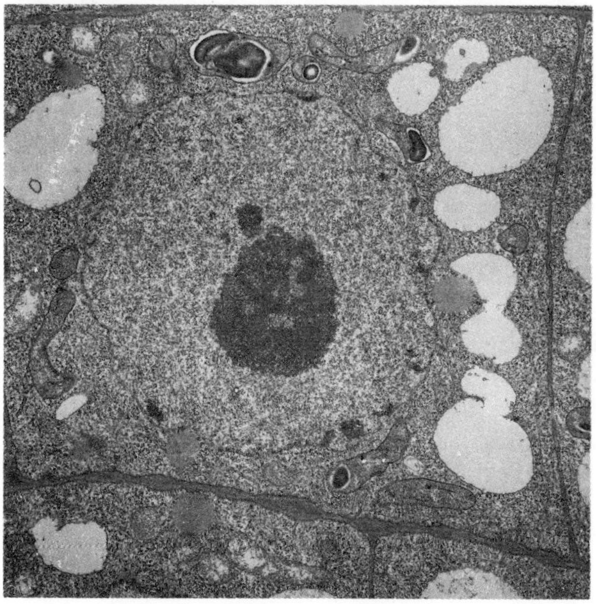

d

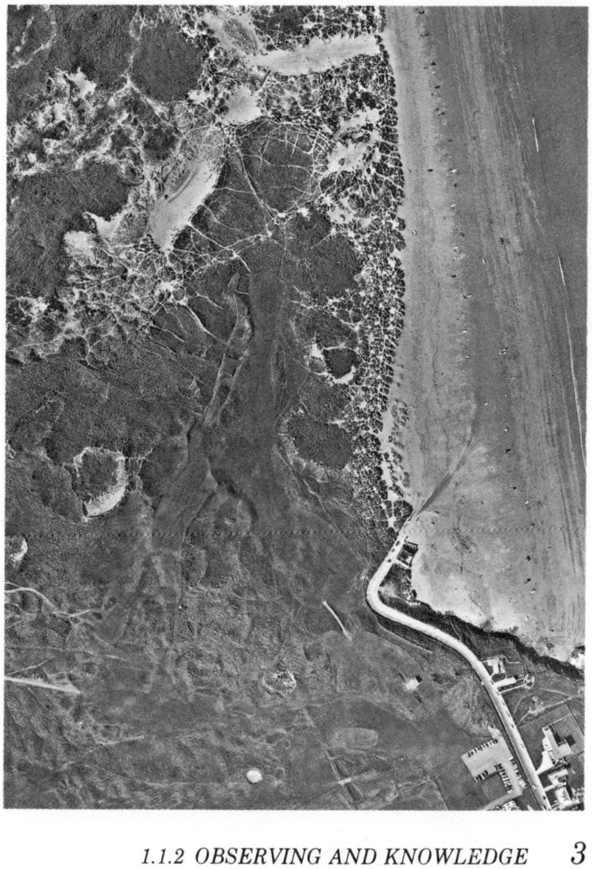

microscope. Photograph (c) was taken with the transmission electron microscope. Photograph (c) was taken with the transmission electron microscope. Photograph (d) is at the opposite end of the scale – it was taken from an aeroplane. Does this information help you to interpret them?

Specialist images such as micrographs, which form the basis of much of biological knowledge, depend on specialist biological knowledge for their interpretation. When you begin to study a new course, you only have a limited knowledge of the subject. Biology involves a considerable amount of observing and you have to build up your knowledge and observation skills in a snowball fashion.

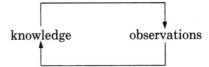

Photograph (c) on Plate 1.1 was taken with a transmission electron microscope. You might have some basic biological knowledge to help you, but you probably will not be able to identify all the structures which are present. They only become meaningful to you when you have been shown similar structures and told what they are.

Let us examine the eye–brain aspect of perception further, by looking at situations where there can be multiple perceptions of the same retinal image. In such cases, the same impulses pass along the optic nerves so that any differences in perception must occur in the visual cortex of the brain.

2 Look at Figure 1.3. There are actually *seven* different ways that we can perceive this object. How many can you 'see'? Keep concentrating on the figure and other perceptions should appear.

Notice that you can only see one perception at a time but you can consciously switch from one to another. Some of these perceptions depend on the fact that we can perceive it in a three-dimensional form although it is only in two dimensions.

The 'Sigmoid Fraud' shown in Figure 1.4 (with thanks to Alan Ternes and Stephen Jay Gould) has two perceptions. Can you see them? As in the previous situation, this one also depends on a three-dimensional perception of a two-dimensional object. Many optical illusions depend on this property.

3 Look at Figure 1.5. Which spider will reach the fly first if they are both travelling at the same speed? After answering, measure the distances with a ruler. What does this particular illusion depend on?

Plate 1.2 contains a few more deceptions for you to look at. Try and work out in each case what causes the illusion.

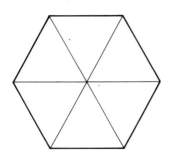

Figure 1.3
Hex figure

Figure 1.4
Sigmoid Fraud

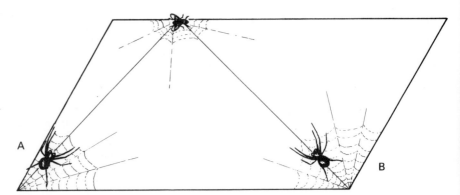

Figure 1.5 Two spiders and a fly

PLATE 1.2

a What do you see where the lines meet?

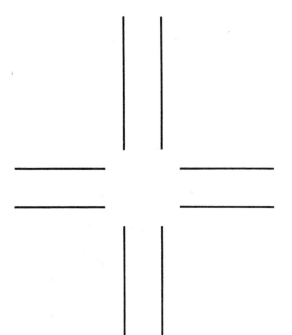

b Which of the lines (A, B, C or D) is continuous with line X?

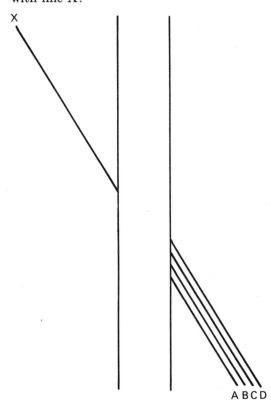

c Is this card sitting up?

d Window frame and shadow.

We can also 'see' things which aren't there at all! In Figure 1.6 below, you will observe a white triangle on top of the rest of the diagram even though it doesn't exist, and if you look closely at Figure 1.7, you will see colours, albeit faintly, although in reality there are only black and white lines.

Figure 1.6 *Subjective triangle*   Figure 1.7 *Subjective colours*

1.1.3 Observing and hypotheses

Observations in biology should always be made with regard to their biological significance. We have already seen that this depends on a certain level of knowledge and competence on the part of the observer. The scientist will often perceive something he or she sees in terms of a hypothesis which is already in mind, what we could call a tentative hypothesis. This starts off a snowball effect whereby he or she will want further information to determine if the tentative hypothesis is correct. The next stage depends very much on the situation, but basically the scientist collects data which is relevant to the problem he or she is considering. The investigator:

(i)   could simply carry out further observations along the same lines as the first investigation in order to obtain more data;

(ii)   could design and carry out an experiment which would test the tentative hypothesis – this could yield further useful data;

(iii)   would be concerned with the type of data obtained. Initially the data might have been qualitative but in order to test the tentative hypothesis he or she needs to obtain more exact data. The observations are quantified, if possible, by counting or measuring the variables under observation;

(iv)   would also consider at this stage what sort of statistical analysis of the data would be appropriate since this affects the type of data that has to be obtained.

Once this procedure has been followed, it should be able to be replicated in exactly the same way by other biologists. The information is then looked at as a guide to the assessment of the initial tentative hypothesis. One of a number of things can happen at this stage:

(i)   The initial hypothesis is found to be incorrect, in which case it has to be abandoned and other ideas or tentative hypotheses are thought up;

(ii)   The initial hypothesis is found to be too general and so more specific hypotheses are formulated;

(iii)   The initial hypothesis is only partly correct so it is modified accordingly;

(iv)   The initial hypothesis is correct in all details and so it is confirmed. This is most unlikely to happen; in reality hypotheses usually require at least some modification.

The following diagram is a model of what happens:

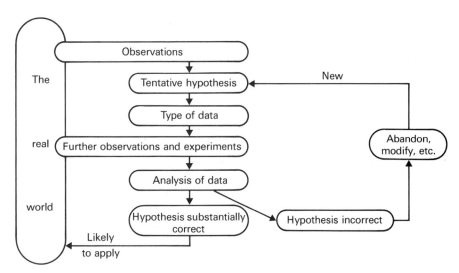

Figure 1.8
The science cycle

As a scientist's store of specialist knowledge increases, not only from his or her own researches but also from others working in the same field, he or she becomes better and better at thinking up hypotheses and designing experiments to test them.

The science cycle illustrates the dynamic nature of science. Every worthwhile hypothesis will give rise to many further hypotheses. The possible solutions to any problem always give rise to further questions.

1.1.4 Observing and investigating

The observations that are made in biology cover an extremely wide range of situations, and the techniques required to deal effectively with them are correspondingly varied. To become competent biologists, you will need to acquire and develop these skills as fully as possible, and the only way to do this is with practice. There is no use in simply reading about observing, you have to do it repeatedly, until you become competent.

Engineers and biologists have much in common, since they both try to relate structure to function. Engineers are primarily concerned with designing structures that will carry out particular functions, while biologists try to find out how a particular structure is related to the function it performs. Often the function of something can be deduced by careful analysis of its structure. In such cases it is necessary to note carefully the similarities and differences between structures, and this can involve recording representations of what is seen by means of accurate drawings. This skill is also used in engineering. The skill of drawing in biology has formed the cornerstone of biological observations over a long period of time and is dealt with in detail in Section 2.

If we want to test initial (and therefore tentative) hypotheses, it is necessary to obtain more exact data. This is achieved by quantifying the variables where possible; indeed one of the main tenets of modern scientific method is *– if it exists it must exist in a certain amount and so we should be able to quantify it*. Quantifying means that we put values to the things that we are observing; we count or we measure the variables in question. Instead of stating that there were 'only a few' or 'many', or that something was 'quite long' or 'very small', we define our observations more precisely. We can obtain discrete or continuous data (see Volume 1, *Dealing with Data*, pages 2 and 3) by counting or measuring respectively.

Quantifying data is necessary because of the huge variability of biological material. Numerical data can be analysed statistically – the results are therefore more meaningful and can often lead to more precise conclusions. These in turn can help in refining the initial hypotheses. Counting and measuring will be looked at in detail in Sections 3 and 4.

1.1.5 Extrasensory receptors

Early work in biology involved recording data which was perceived using our senses of sight, hearing and touch, etc. Gradually however, techniques and apparatus were developed that could detect phenomena beyond our normal sensory experience.

Much effort and ingenuity has particularly been expended in trying to extend our range of vision. It has been possible to view smaller and smaller objects with the aid of microscopes, and larger and larger objects with remote viewing techniques. Many devices have been specifically designed to allow us to see data which would normally be received by another of our senses. As an example, you could place your hand on someone's forehead to find out if he had a temperature, and you might be able to say yes or no. It is better to place a clinical thermometer under the tongue to measure their temperature, but this involves reading the temperature accurately, which can be difficult. Recently, digital thermometers have been developed which give a clear readout of the temperature, making the process less prone to human error.

Many of these 'extrasensory receptors' are outside our normal experience and the information they produce is often difficult to understand. As in Section 1.1.2, it takes time and effort to make it meaningful. Much of this book uses techniques that have been developed to extend the use of our senses, particularly sight. It is important during investigations that the appropriate techniques are used, at a proper level of accuracy. These devices help to record observations more accurately, using drawing, photography, counting or measuring.

1.1.6 Microscopy

Much of the progress in biology over the years has been due to the microscope, an instrument which is very closely linked with the subject.

The development of the microscope

The first microscopes were developed during the seventeenth century, although glass lenses have been in use since the mid-thirteenth century.

During the seventeenth and early eighteenth centuries, Anton van Leeuwenhoek in the Netherlands and Robert Hooke in England worked independently to develop microscopes. Van Leeuwenhoek was the first to demonstrate the existence of bacteria, red blood cells, and other biological structures.

Throughout the eighteenth century, many attempts were made to minimise the imperfections in the images produced by lenses. Basically there are two types of imperfection:

Figure 1.9
Leeuwenhoek

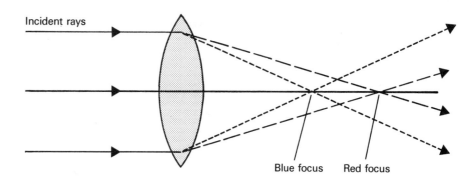

Figure 1.10
a Chromatic aberration

Incident rays

Blue focus   Red focus

(i) **Chromatic aberration** – the inability of a lens to bring light of different colours to focus at the same point;

(ii) **Spherical aberration** – the failure of a lens to focus rays of light passing through the edge of the lens on to the same point as rays that pass through the centre.

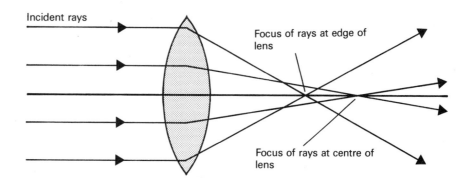

Incident rays

Focus of rays at edge of lens

Focus of rays at centre of lens

b Spherical aberration

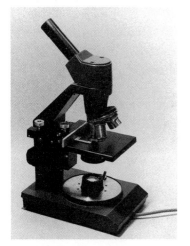

Figure 1.11
Compound microscope

Ernst Abbé (joint founder of the Zeiss optical company at Jena) helped to solve these problems and advance microscopy. He realised that even in the most optically correct lenses the theoretical point of light can never be attained, as it always spreads to become a circle. He proposed that this was not due to imperfect microscope design, but depended on the fundamental nature of light.

By this time brass was being used to construct microscopes, which allowed very accurate machining, so that for example, fine focusing mechanisms could be incorporated.

By the nineteenth century, the microscope had acquired its modern form and had become a precision instrument. Most subsequent developments have been improvements to the lighting system used.

Bright-field microscopy

This is the basic microscope technique, in which a bright light-source is directed through the specimen. Looking down the microscope, a bright circle of light can be seen – this is the **field of view**. When a thin slice of the specimen is placed in the field of view, different regions of the specimen absorb and transmit different colours and amounts of light. This happens particularly if the specimen has been stained.

Image formation

The simplest way to consider the optics of the compound microscope is to imagine that it consists of two simple convex (converging) lenses. One (OO) is near the object, and is called the **objective lens**. The other (EE) is close to the eye and is known as the **eyepiece lens** (see Figure 1.12).

Light from an object (PQ) forms an image (P'Q') when it passes through a convex (converging) lens. This is much the same as the way an image is formed in a camera – the image is inverted. The difference in size between the object PQ and the image P'Q' depends on the magnifying power of the lens and the distance of the object from the lens.

The second lens (EE) is used to view the image P'Q' in the same way as a magnifying glass is used to read small print. The final image P''Q'' is therefore still upside-down.

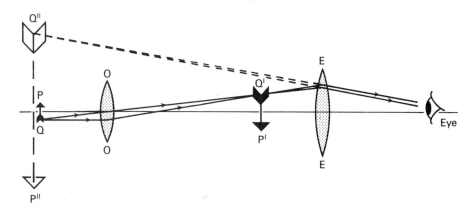

Figure 1.12
Image formation in the compound microscope

The actual path of the light rays through the compound microscope is in fact much more complicated than shown in the diagram, which applies only to infinitely thin lenses and assumes that they are capable of forming a point image from a point source. As previously described, most lenses have defects and aberrations, and in order to correct these, combinations of lenses of different curvatures and materials are used.

Magnification and resolution

It is not true that by using higher and higher magnifying power, the smallest details may be seen using a microscope.

The magnification of a lens is the number of times the image is linearly larger than the object. Referring to Figure 1.12, we can see that the object (PQ) is magnified first by the objective lens (OO), to produce the enlarged image (P'Q'). This is in turn magnified by the eyepiece lens (EE), to give the final image (P''Q''). The total magnification of a compound microscope is calculated as follows:

$$\text{total magnification} = \frac{\text{magnifying power of}}{\text{objective lens}} \times \frac{\text{magnifying power of}}{\text{eyepiece lens}}$$

The magnifying powers of the objective and eyepiece lenses are normally clearly engraved on the barrel of the lens, for example ×20, ×40, etc. To assist with identification, the objective lenses are also usually colour-coded.

The magnification produced by a lens depends on the distance between the lens and the image. Therefore, when two or more lenses are combined, the distance between the lenses is important. Most microscope manufacturers have standardised these distances so that the lenses are 160 mm apart, and

Figure 1.13
Two objective lenses

the final image is 210 mm from the eye. This means that the total magnification can be worked out as shown on the previous page.

A second set of figures is also engraved on the barrel of an objective lens (see Figure 1.13) – this is the ***numerical aperture (NA)*** of the lens and this value usually ranges from about 0.1 to around 1.4. The numerical aperture is inversely related both to the diameter and to the focal length of the lens.

1 There are two lenses of magnification ×100. One (A) has a diameter of 30 mm and a focal length of 25 mm; the other (B), has a diameter of 5 mm and a focal length of 10 mm. Which lens would have the highest NA value?

This shows that the higher the magnification of the objective lens, the smaller the lens.

Figure 1.14
Objective lenses viewed from the side and from below, showing magnification and diameter

The importance of the numerical aperture in microscopy is that it determines the ability of the lens to show fine detail, the same property identified by Abbé as ***resolution***. This is defined as the minimum distance between two points which allows them to be distinguished as separate entities using a microscope. The relationship between numerical aperture and resolution is:

$$R = \frac{0.61\lambda}{\text{NA}}$$

where R = the minimum resolvable distance
and λ = the wavelength of the light used

So the smaller the value of R, the better is the resolving power of the microscope, and the finer the detail it can show.

2 Which of the two lenses (A or B) in Question 1 would have the best resolving power?

3 Given that the wavelength of green light (in the middle of the visible spectrum) is 550 nanometres, calculate the minimum resolvable distance of an objective lens of magnification ×40 and NA of 0.65.

The numerical aperture also varies according to the substance that the light passes through before entering the lens. This is one of the reasons why very high power objectives (×100 and greater) have the gap between them and the slide filled with a special oil. This is an **oil-immersion** objective.

4 Calculate the minimum resolvable distance of an oil-immersion objective of magnification ×100 and NA of 1.25.

5 Resolution is also affected by the wavelength of the light used. Calculate the minimum resolvable distance of a ×40 objective with NA of 0.65 using red light of wavelength 650 nm.

6 Repeat Question 5, this time using blue light of wavelength 450 nm.

7 What effect does the wavelength of the light have on resolution?

Diatoms have been used since the mid-eighteenth century as a means of testing microscopes for resolution and general performance. The case of a diatom is highly sculptured with patterns of dots in lines. In some species the dots are about four per micrometer, 0.25 μm apart, which is just about at the limit of resolution for the light microscope.

The purpose of the microscope is therefore not simply to magnify objects, but also to resolve the detail of their structure and then magnify the image to a size which is comfortably visible to the human eye. Bear in mind that the calculations in Questions 3 to 6 are theoretical. Because of limitations due to the design and construction of the microscopes, the actual resolution of the best bright-field microscopes is likely to be about 0.25 μm, while those used in schools, under non-ideal conditions, have poorer resolution.

Other types of light microscope

Most of the developments in microscopy during the nineteenth and twentieth centuries have been attempts to improve resolution. Greater magnification has been required to resolve smaller objects, and make them visible to the human eye.

Good resolution is normally only obtained with thin and virtually transparent specimens which require no manipulation. Many biological specimens are however thick and opaque, and so it is necessary to carry out work on them while observing through the microscope. This necessitates the ability to judge distance and to have a three-dimensional view of the specimen, that is, a **stereoscopic** view. Stereoscopic vision is possible in humans because the eyes are situated at the front of the head, sufficiently far apart so that each eye sees the object from a slightly different angle. When reading, the eyes are normally inclined towards the nose at about 15°. A stereomicroscope consists of two microscopes inclined at an angle of about 30° so that the image observed by each eye differs in the same way as normal vision. The stereomicroscope is widely used throughout the natural sciences: it allows reasonably fine detail to be observed without the specimen having to be either very thin, transparent, killed or sectioned. Although up to ×200 magnification is possible, between ×5 and ×100 are more typical and useful.

8 What sorts of procedures would the stereomicroscope be useful for?

Although all stereomicroscopes are binocular the reverse does not hold true. A true stereomicroscope has two eyepieces **and** two objective lenses – this should not be confused with monocular microscopes which have a single objective lens and a binocular eyepiece.

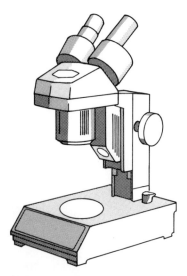

Figure 1.15
Stereomicroscope

When light passes through a specimen, several changes in the light take place which form the image observed by the eye.

Changes 1 to 5 shown in Figure 1.16 are most important in **bright-field microscopy**. Many of the procedures used in the preparation of specimens, particularly sectioning, squashing, smearing and clearing, increase the penetration of light through the specimen. A wide range of stains have been developed which enhance and create colours; these help in the recognition of different chemical components in protoplasm, cell types and tissues. The remaining changes have each been made use of by the development of special microscopes or lighting techniques, particularly for observing living cells without fixing or staining. In such cases there are no artefacts which are caused by chemical treatments and it is possible to view continuous processes, for example cell division in their entirety. Change 6, caused by the alignment of long-chain molecules in cytoplasm, has been demonstrated using polarised light. **Polarising microscopes** have been most useful in the study of the cytoskeleton and the spindle of dividing cells.

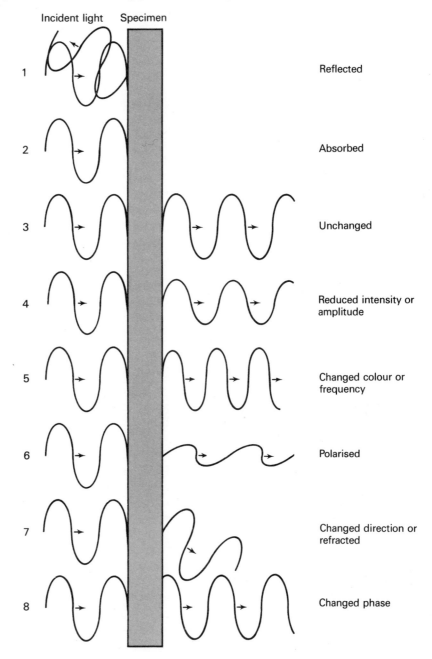

Figure 1.16
Effect of light passing through a specimen. These effects occur in combination and in varying degrees, although they are separated in the diagram for clarity

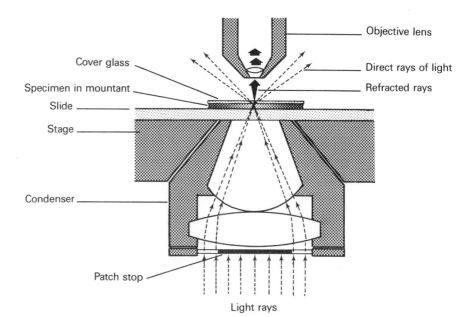

Figure 1.17
Dark-field illumination

Refraction (bending of light), change 7 in Figure 1.16, results from slight differences in the refractive index between the specimen, or regions of it, and the glass of the slide and the mountant. One use of this characteristic has been **dark-field illumination**.

As shown in Figure 1.17, the name comes from the fact that no light is allowed to pass directly through the specimen and objective lens from the light-source, so that the background appears dark. However, light is allowed to strike the specimen at an angle. Those parts of the specimen which refract the light cause light to pass into the objective lens and hence they appear bright on a dark background.

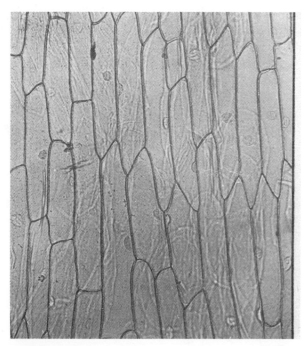

a Bright-field (×120)

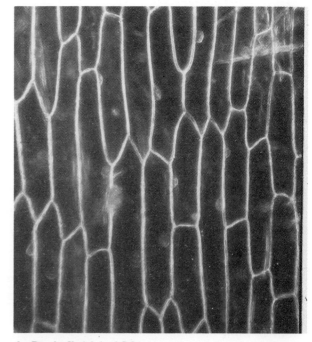

b Dark-field (×120)

Figure 1.18
Onion epidermis

Changes in the phase of light as it passes through a specimen have also been used extensively in biological research. Although the detailed theory of **phase-contrast microscopy** is beyond the scope of this book, the principles worked out by Zemicke in 1940 are as follows:

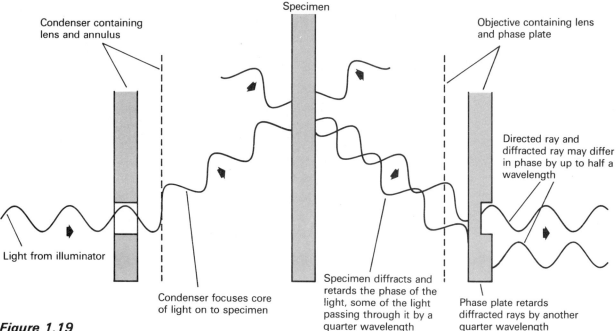

Condenser containing
lens and annulus

Specimen

Objective containing lens
and phase plate

Light from illuminator

Condenser focuses core
of light on to specimen

Specimen diffracts and
retards the phase of the
light, some of the light
passing through it by a
quarter wavelength

Directed ray and
diffracted ray may differ
in phase by up to half a
wavelength

Phase plate retards
diffracted rays by another
quarter wavelength

Figure 1.19
**Light path in a phase-
contrast microscope**

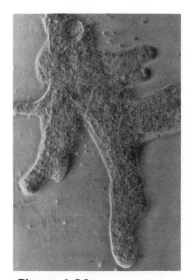

Figure 1.20
**Phase-contrast specimen
of Amoeba**

Contained within a phase-contrast condenser is a special glass plate which allows only a circle of light to pass through when focused on the specimen. This produces a hollow cone of light. When there is no specimen present this light passes directly into the objective lens and through another special glass plate, the **phase plate**. This plate is made of clear glass with a ring-shaped slot cut into its surface, and is positioned so that the direct light passes through the ring of thinner glass. When a specimen is present, some of the direct rays will be changed in phase and direction (diffracted), passing into the objective lens and through the thicker region of the phase plate. The different thicknesses of glass passed through by the direct and diffracted light effectively increase any phase difference caused by the specimen.

Although the human eye cannot directly detect phase changes in normal light, in the phase-contrast microscope, contrast between direct and diffracted rays is achieved. The eye sees patterns of light and dark in the specimen because the direct and diffracted light waves reinforce each other to a greater or lesser extent. Similar principles using two light sources produce coloured interference patterns in an **interference microscope**.

Another series of developments are based on the relationship between the resolving power, the wavelength of the light and the numerical aperture (see page 11). Reduced values of R (minimum resolvable distance) may be achieved by either increasing the NA of the lenses used, or by reducing the

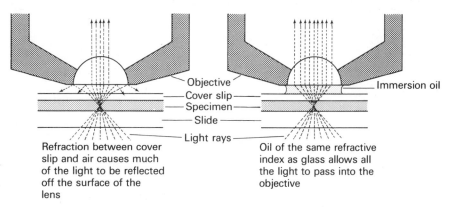

Objective
Cover slip
Specimen
Slide
Light rays
Immersion oil

Refraction between cover
slip and air causes much
of the light to be reflected
off the surface of the
lens

Oil of the same refractive
index as glass allows all
the light to pass into the
objective

Figure 1.21
**Light path in dry and oil-
immersion objectives**

wavelength of the light. Increasing the NA beyond 1.4 means having to use smaller and smaller lenses, using oil-immersion lenses and condensers (see Figure 1.21), and reducing the working distance.

Reducing the wavelength of the light involves using blue filters with ordinary light sources, and developing ultraviolet and X-ray wavelengths as alternative light sources. The latter two have limited use in biology since both result in the death of living cells. Other practical problems mean that the resolution achieved is not as good as the theoretical values – approximately 0.1 μm with UV light.

Electron microscopes

Improvements in resolution became possible in the 1920s and 1930s when it was discovered that electromagnetic fields could focus electrons. This made the construction of the **electron microscope (EM)** possible.

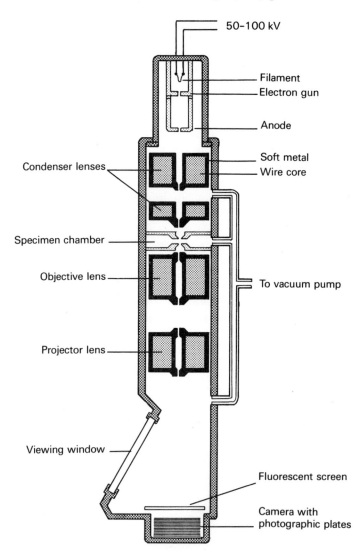

**Figure 1.22
Construction of the
electron microscope**

In its simplest form, the electron microscope has the same basic design as the compound optical microscope, described previously. Any differences are due to the properties of electrons. The most important change necessary when using electrons is the removal of all the air from the tube of the microscope, because electrons are easily stopped by any molecules in their path. In optical microscopes the light is produced by heating a filament with electricity. To produce electrons, electricity is used to heat a filament of tungsten to over

3000 °C. The negatively charged electrons are then attracted a short distance to a positively charged anode plate which has a small, centrally positioned hole. High voltages (50–100 kV) cause many of the electrons to pass through the hole in the anode and down the length of the microscope. The density of the electron beam is further increased by the inclusion of a negative cathode shield between the filament and the positive anode. Collectively this part of the microscope is known as the *electron gun* – similar structures are found in cathode-ray and television tubes. Increasing the accelerating voltage used in the electron gun reduces the wavelength of the electrons and thus improves the resolution of the microscope.

EM lenses consist of several thousand cords of wire encased in soft iron. A current of about 1 A passes through the wire and produces a focal length of a few millimetres. Small increases of current in the coil reduce the focal length of the lens. The overall magnification produced is controlled by the current in the projector lens.

The structures described above are common to all EMs but in the last few decades two distinct designs have been developed to investigate different types of specimen, namely the **transmission electron microscope (TEM)** and the **scanning electron microscope (SEM).**

**Figure 1.23
Preparation of sections
for the TEM**

(a) Fixing – pieces of specimen are immersed in chemicals such as ethyl alcohol, ethanoic acid, formaldehyde or osmium tetroxide, which chemically bind the structures of the cells in a fixed position and generally harden the tissue

(b) Dehydrating – fixed specimens are placed in a series of alcohols or acetones of increasing concentration which remove all water from the cells

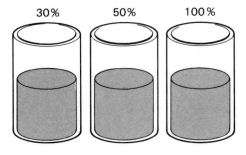

30% 50% 100%

(c) Embedding – specimens are placed in a resin which is then baked in an oven until hard

(d) Sectioning – the specimen embedded in resin is trimmed and placed in the chuck of an ultramicrotome, which moves up and down past the cutting edge of a glass knife. Sections float off into a small amount of water and are retained by a piece of tape.

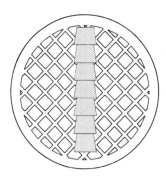

(e) Mounting – the sections are lifted from the microtome on to a small copper grid

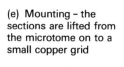

(f) Staining – the grids with sections on the underside are placed on top of single drops of heavy metal stains, such as lead citrate and uranyl acetate

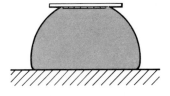

The transmission electron microscope (TEM)

TEMs are very similar to optical microscopes, with the specimens being smeared or ultra-thinly sectioned (between 0.05 and 0.1 μm) and mounted on delicate copper grids about 2 mm in diameter (see Figure 1.23 on the previous page). Nearly all the electron beam passes through these ultra-thin sections because they contain only molecules of low atomic number, for example carbon, hydrogen, oxygen and nitrogen, which stop or scatter few of the electrons. Very little detail of the structure of the specimens is therefore discernable.

To provide the necessary contrast the specimen is treated with chemicals containing heavy metals: osmium tetroxide during fixation, lead citrate and uranyl acetate during staining. These molecules become attached to particular structures and chemicals in cells, either blocking or scattering electrons when they strike them.

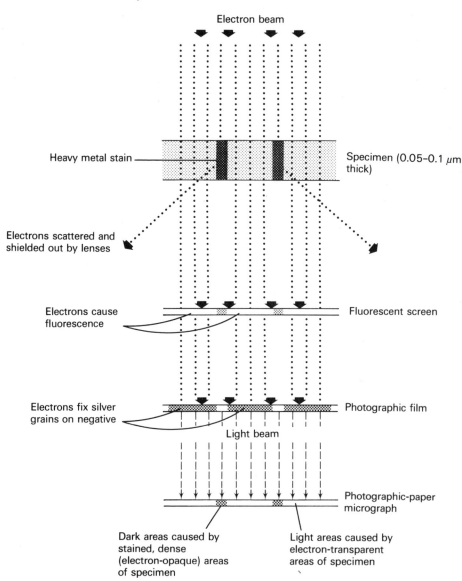

Figure 1.24
Image formation in the TEM

The remaining electrons continue down the microscope tube and are made visible on striking a screen covered by a chemical such as zinc sulphide. This gives a greenish-grey image. The screen can be observed through a window at the bottom of the tube and precise focusing is carried out using low-powered binoculars. For a permanent record, the screen is moved to one side and the electrons are allowed to fall on to a photographic plate which

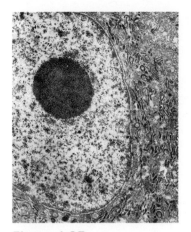

Figure 1.25
TEM tegument cell of Opisthioglyphe ranae (×8400)

when developed and printed gives an ***electron micrograph***. Dark areas on a micrograph correspond to regions of heavy metal staining in the specimen which are electron-opaque as they do not let electrons pass through. Light areas of the micrograph are electron-transparent since they let electrons pass through.

The magnification used in TEMs ranges from ×1000 to ×200 000, with most biological work being carried out well below the maximum because the nature of the material and the methods of preparation limit the resolution. Published micrographs occasionally have magnifications in excess of ×200 000. This is because the final print magnification is the microscopic magnification multiplied by the number of times the negative is enlarged during the photographic printing process. Using TEMs to analyse crystals has allowed researchers to separate points of the crystalline lattice which are slightly more than 0.2 nm apart, while the limit of resolution of biological material is around 1 nm.

9 How much better is the resolution of biological material by a TEM compared with
  (i) an optical microscope, using green light?
  (ii) an ultraviolet microscope?

Two alternative methods of obtaining contrast have been developed for use on very small objects such as viruses, microtubules and fragments of DNA isolated from cytoplasm. Specimens of these dimensions are too small to be carried on a grid alone and so a thin plastic film is used to support them.

The first of the techniques to be developed involves placing the grid of specimens in a vacuum chamber. Above and to one side of the specimen is a metal filament (usually gold or platinum) which, when heated, releases atoms in straight lines towards the specimens. Heavy metal therefore accumulates more on one side of each particle than on the other. When placed in the TEM, the thick layer of metal appears dark on the screen, while the side with no metal appears light. For this reason the technique is known as ***shadowing***.

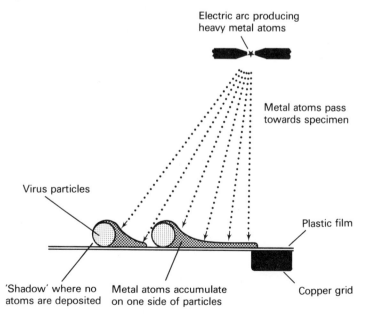

Figure 1.26
Shadowing

Knowledge of the distance and the angle between the electric arc and the grid, and the size of the shadow allows the size of the particles to be determined

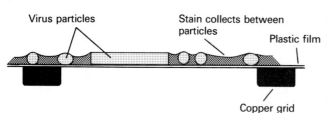

Figure 1.27   *a Negative staining*

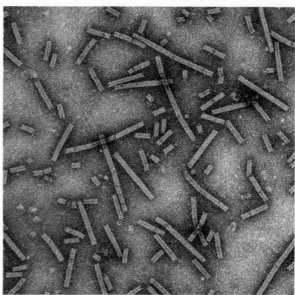

b Potato mosaic virus

The second technique, known as **negative staining**, is simpler. Specimens, mounted as previously described, are covered with a solution of a heavy metal stain. After the solvent evaporates, the stain forms a continuous layer around the specimen, which appears light against the dark background of the stain.

Both of these techniques have been very important in providing accurate data about the size and shape of cell fragments.

A more recent adaptation of shadowing, called **freeze-etching**, has provided strong confirmation that the chemicals used in preparing TEM sections have minimal effect on the cell structures. In this case the cellular material is fast-frozen by plunging it into liquid nitrogen at −210 °C. It is then fractured while frozen using a razor blade, and it splits along the weakest parts of the cell structures – the membranes. To further expose these membranes, water is sublimed from the frozen specimen while in a vacuum. This stage is called etching. A replica is then made by releasing carbon molecules on to the specimen from above in the same way as for shadowing. This replica is necessary as it is not possible to use such a specimen in the TEM. Finally the replica is shadowed with heavy metal as described previously (see Figure 1.26).

The scanning electron microscope (SEM)

This simpler type of electron microscope was developed in the early 1960s. Outwardly, it appears similar to the TEM and it has many common elements internally, but it is based on a fundamentally different principle of image-formation.

Both types of electron microscope generate electrons from an electron gun, but SEMs use a much lower accelerating voltage (1–30 kV). The SEM beam of electrons passes through several electromagnetic lenses in the evacuated tube, similar to the condenser lenses of the TEM, but produces a much finer beam of approximately 5 nm in diameter. The dimensions of this beam are critical to the eventual resolution of the instrument. An additional coil moves the beam back and forth across the specimen area at the bottom of the tube.

The preparation of specimens is usually less complicated than for TEMs. Some structurally rigid items such as pollen grains and shells need no chemical treatment, while others require fixation and dehydration. Very delicate methods have been devised for cells and tissues to reduce damage. Freeze-fracturing or freeze-drying (fast-freezing followed by the sublimation

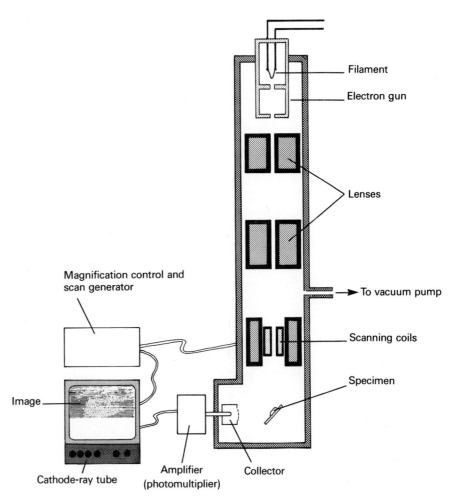

Figure 1.28
SEM structure

of water under vacuum before embedding) may also be employed. All specimens are finally coated with a thin layer of conducting material, often carbon, and occasionally gold, using the same technique as for replica-production described earlier.

The electrons passing down the tube are known as primary electrons and the specimen is usually held at an angle to these, although the position can be changed. The primary electrons collide with the molecules making up the surface of the specimen, resulting in the emission of X-rays, light and most importantly more electrons known as secondary electrons. Most SEMs produce an image based on these secondary electrons, but some instruments can work with the other emissions. The secondary electrons are attracted to a positively charged collector which also moves back and forth across the specimen area in synchrony with the primary electron beam. Within the collector, electrons are accelerated, converted into light in a scintillator and then into an electric current by a photomultiplier. This current is used to change the intensity of a cathode-ray beam. A picture is built up in the same way as in a television set, with the cathode-ray beam moving back and forth across a fluorescent screen. Photographs can easily be taken of the image.

The parts of the specimen which produce the most collectable secondary electrons therefore appear the brightest on the cathode-ray image. These are usually the sloping sides nearest the collector, so the overall impression given is that a light is being shone on the specimen from one side. Unlike a photograph, all parts of the image are in focus, giving a remarkable three-dimensional appearance to SEM micrographs.

10 Write down what you think the SEM micrographs are in Plates 1.3 and 1.4.

PLATE 1.3

a

b

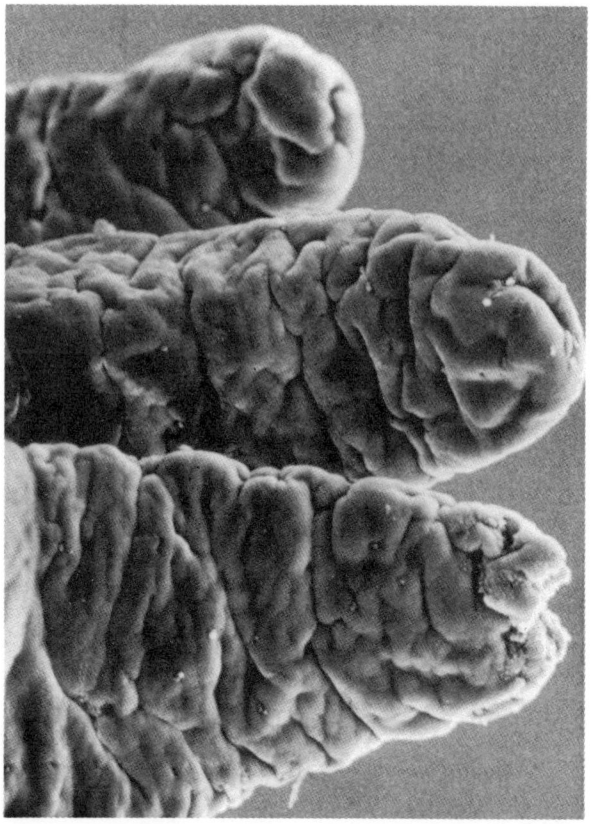

c

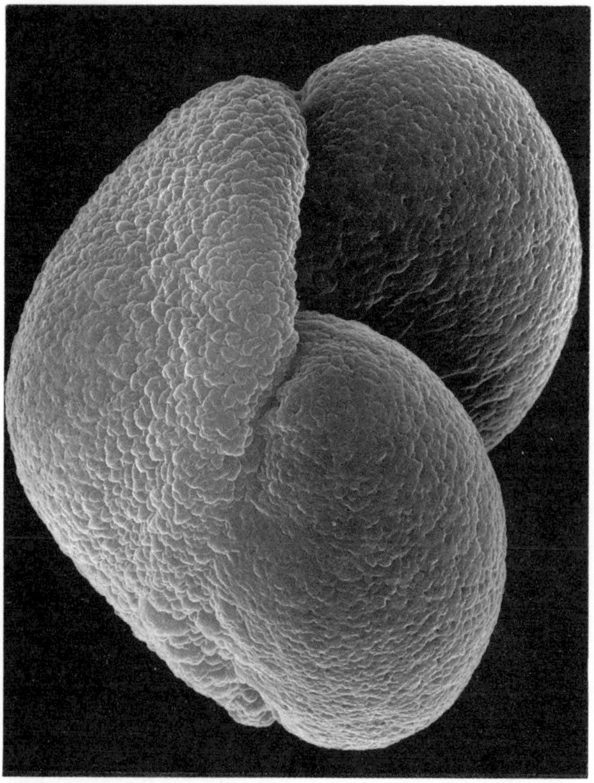

d

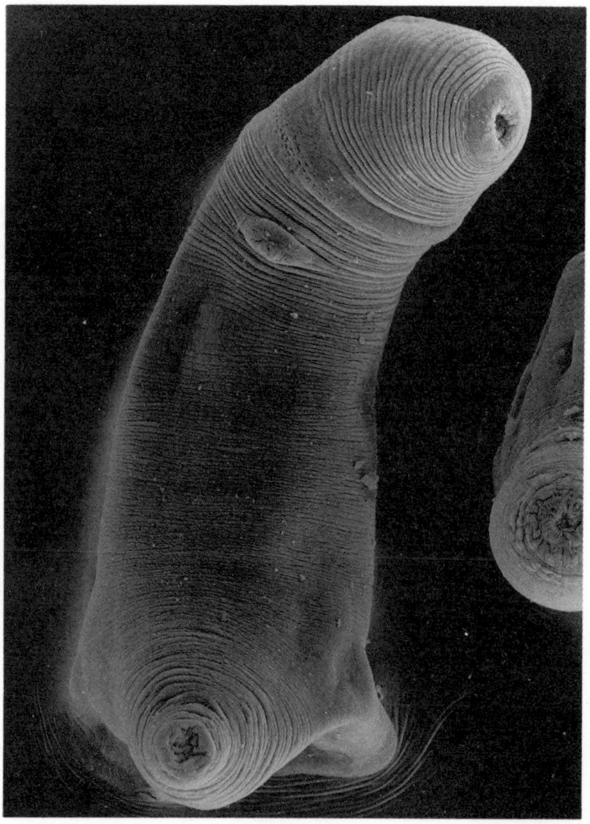

PLATE 1.4

a

b

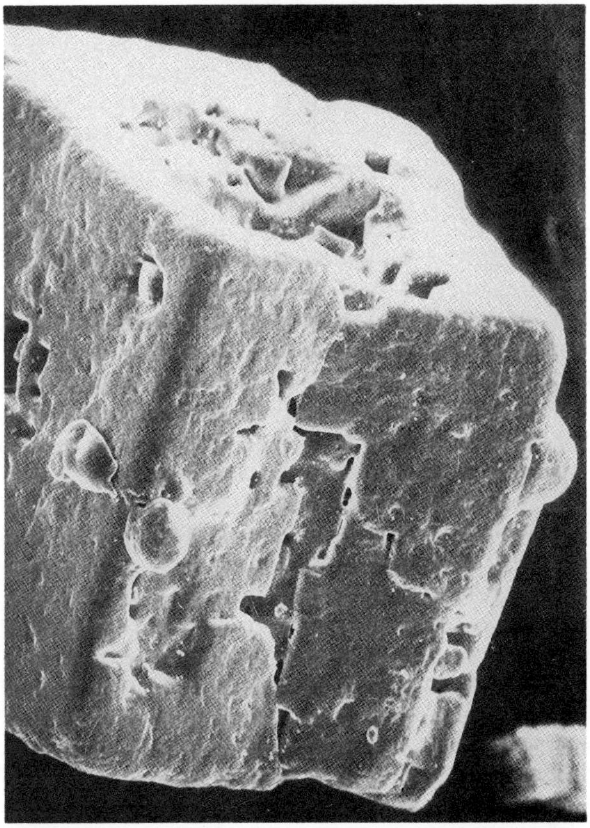

c

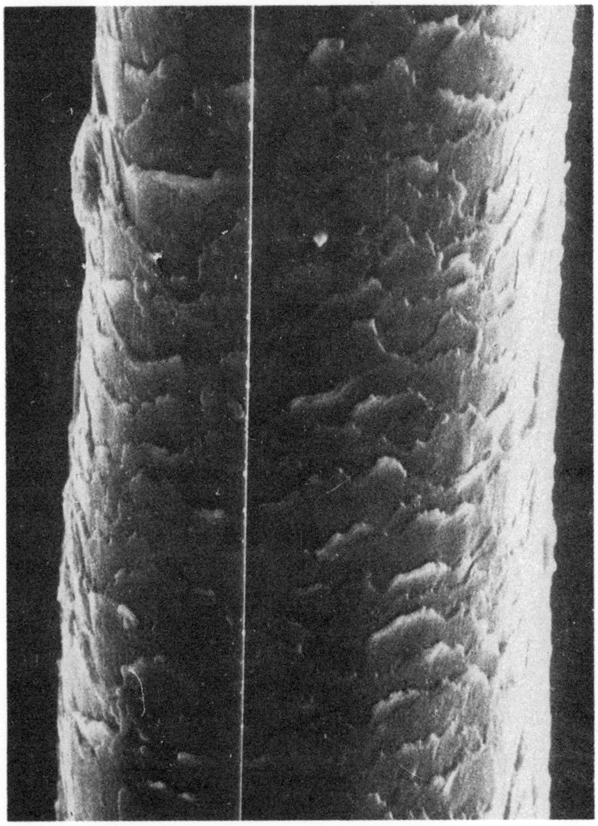

d

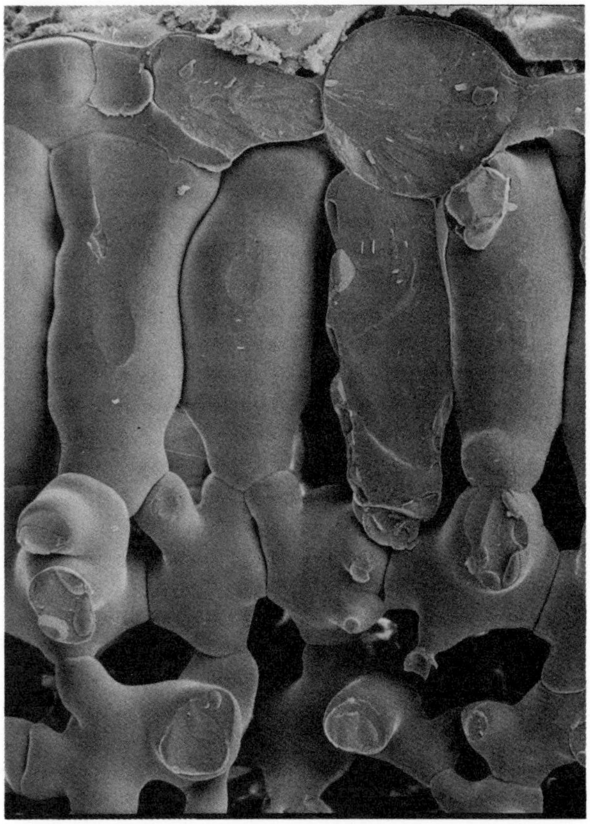

The SEM is much more flexible and simple to operate than the TEM, but it does not achieve the same level of resolution – this depends on a number of factors, for example the diameter of the beam, the nature of the specimen, and the scanning speed, as well as the density of lines making up the final image. A minimum resolvable distance of about 10 nm is achievable.

The range of magnifications available is extensive, ×10 to ×100 000, overlapping hand lenses at the lower end and TEMs at the upper limit of magnification.

1.1.7 Remote sensing

Remote sensing can be described simply as 'the distant observing of objects'. It is an ideal technique for monitoring the earth's resources when vast tracts of land, often inhospitable, must be surveyed within short time spans.

Instruments have been developed to reduce extremely large objects to a size which lies within our viewing range. The further we are from the object under observation, the smaller it looks, but the greater will be the field of view. As the distance increases however, the resolution becomes poorer.

Passive sensing

In passive sensing the sensor detects energy waves which arise naturally from the object or are reflected by it. These energy waves, and thus the nature of the radiation, is defined by their wavelength. Different parts of the electromagnetic spectrum can be detected by different sensing devices.

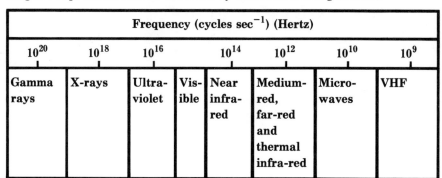

Frequency (cycles sec^{-1}) (Hertz)							
10^{20}	10^{18}	10^{16}	10^{14}	10^{12}	10^{10}	10^{9}	
Gamma rays	X-rays	Ultra-violet	Vis-ible	Near infra-red	Medium-red, far-red and thermal infra-red	Micro-waves	VHF

Figure 1.29
Electromagnetic spectrum and sensors used for remote sensing

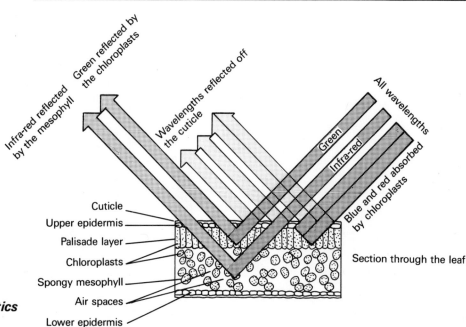

Infra-red reflected by the mesophyll

Green reflected by the chloroplasts

Wavelengths reflected off the cuticle

Green

Infra-red

All wavelengths

Blue and red absorbed by chloroplasts

Section through the leaf

Cuticle
Upper epidermis
Palisade layer
Chloroplasts
Spongy mesophyll
Air spaces
Lower epidermis

Figure 1.30
Reflectance characteristics of a leaf

The main part of a plant is the leaf, which manufactures food using part of the electromagnetic spectrum as its energy source. Every plant species has a unique leaf structure and contains particular amounts of the various photosynthetic pigments. Each species has a characteristic absorption spectrum consisting of the wavelengths of light it absorbs (see Figure 1.30). From this it follows that those wavelengths that are reflected by the leaves will also be characteristic for that species, thus forming what is known as a **reflectance signature**.

These reflected wavelengths consist mainly of infra-red waves since about 80 per cent of the infra-red radiation falling on a leaf is reflected. These reflectance signatures can be used to monitor the presence of particular tree and crop species.

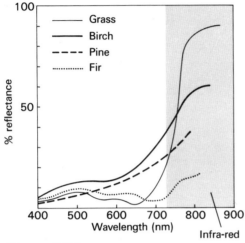

Figure 1.31
Reflectance signatures of various plant species

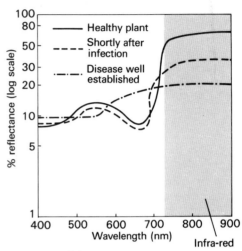

Figure 1.32
Reflectance values of diseased and healthy plants

The reflectance spectrum is affected by the condition of the plant. If the plant is under stress then the reflectance signature is altered and measurements of it can detect disease, drought or pollution. The infra-red waves are again particularly important and these can be used for the early diagnosis of such conditions.

We can therefore use remote sensing techniques to assess the quality, quantity and condition of essential renewable resources such as crops and forests on a vast scale.

The sensors have to be placed at suitable altitudes above the earth's surface. All sorts of devices have been used to house them, including balloons, sailplanes, gliders, microlights, helicopters, aircraft of all types, and satellites. These have given altitudes ranging from a few metres up to 36 000 km. Aerial photography and satellite imagery are the two sources from which most of the information concerning the earth's biological resources is derived.

Aerial photography
Aircraft are the most widely used method for aerial photography since they give plenty of scope for many forms of remote sensing up to about 20 km. Most of the world's natural surface vegetation has now been recorded in this way. One of the most detailed and comprehensive records of changing vegetation is the photo-coverage of the British Isles by the RAF in 1980–1 at a scale of 1 : 50 000. Aerial photographs taken at an angle are particularly good for counting animals, but it is necessary to use vertical aerial photography for systematic ecological mapping or for any purpose which involves measuring (photogrammetry).

Figure 1.33
Vertical aerial photograph

Vertical photographs are also essential for stereo-viewing, which gives the observer the impression of depth. The higher the aircraft, the greater will be the distance between the sensor and the earth's surface and so the greater the scale ratio will be.

Aerial photography is used widely in surveying. The particular scale used and thus the altitude from which the photographs are taken depends on the requirements of the survey. The type of film used is also important.

Survey type	Biological significance	Scale	Film type
Land-use	Investigating natural and induced vegetation; forests; scrub, etc.	1 : 90 000	Black and white; colour; infra-red
Agricultural	Crop identification, maturity and production estimates	1 : 10 000	Black and white; colour
Forest	Finding tree types; areas; inventories	1 : 10 000– 1 : 50 000	Black and white; colour
Animals	Population estimates; breeding areas	1 : 500– 1 : 1000	Black and white; colour
Pathology	Investigating plant diseases and pests – viral, fungal, bacterial; insect damage	1 : 800– 1 : 20 000 1 : 2500– 1 : 8000	Black and white; colour; infra-red
Ecological mapping		1 : 200– 1 : 20 000	Black and white; colour; infra-red

Figure 1.34
Photograph type and appropriate film type

Light aircraft and helicopters have been found to be very efficient in locating fish schools, but only for those pelagic forms which live in the upper zones of the ocean. Present-day commercial applications of aerial sensing are widespread, in fact fishing for species such as menhaden, and to some extent tuna, is very dependent on remote sensing techniques.

When photographing fish schools, near infra-red (wavelengths just outside the visible range) film is used since, like plants, fish schools have different reflectance signatures which can be used to determine the species. Photographs can therefore not only tell the location of the fish school but also its size (biomass) and its quality (species).

Satellite images

The advent of the space age in the last few decades has brought unrivalled opportunities for remote sensing which now provides extensive cover of the earth's surface and the plant and animal life that is present on it. For biological studies the most important satellites are unmanned and have inclined orbits which are relatively close to the surface of the earth, so that areas of the surface that have visible vegetation are well scanned.

Environmental monitoring began with the launch in 1972 of the American ERTS-1 (Earth Resources Technology Satellite) which was later named Landsat 1; this was followed by Landsats 2, 3, 4 and 5. The French Spot-1 was launched in 1985 and the European Space Agency launched their own ERS-1 in 1987.

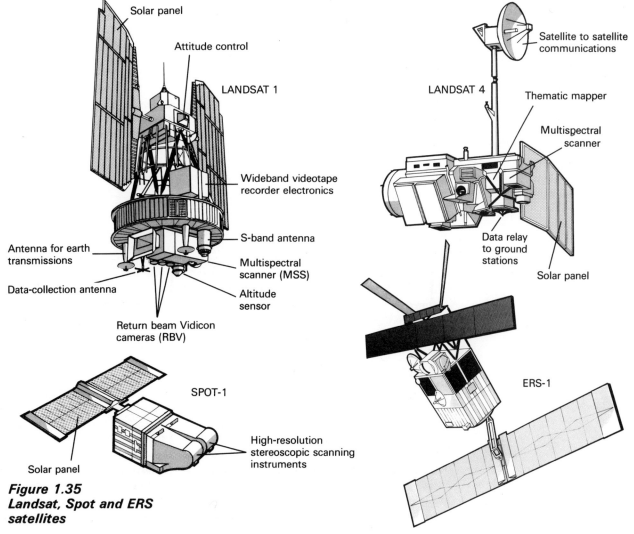

Figure 1.35
Landsat, Spot and ERS satellites

These satellites have specialised sensors so that they can provide unique opportunities for biological studies. Landsats 1, 2 and 3 posses multispectral scanners (MSS) which are capable of measuring reflected electromagnetic radiations over a range of wavelengths. More recent sensors are sensitive to a wider range of wavelengths or have improved resolution. Landsats 4 and 5 have thematic mappers which cover seven bands of the spectrum with a resolution of 30 m and in the future could provide the possibility of stereo-viewing from space; the Spot satellite covers only three spectral bands but the resolution is down to 10 m. The Landsats also carry a bank of Vidicon cameras which sense different spectral bands and these produce video recordings which are transmitted to earth receiving-stations and the images enhanced (improved). The ability of raw satellite images to be enhanced using a variety of procedures is one of their chief attributes since it enables the maximum available information to be extracted. Computers play an important role in *contrast stretching* where areas differing only slightly in shade can be made to contrast more strongly.

Animals pose quite different problems from plants. They are relatively small, they often occur singly, and they move about. To overcome these problems, techniques have been developed whereby variables are measured which correlate strongly with the presence of certain animals, for instance, the mapping of particular habits of certain species is used for planning and management for their protection.

Satellite imagery has been used successfully in helping to control pests like locusts. Locusts need rain for egg-laying and it is also required for the growth of vegetation. Rainfall occurrence can be confirmed by colour changes on Landsat images due to the moist soil and the emerging vegetation. High-risk areas can therefore be identified and teams on the ground directed to them so that the locusts can be controlled.

In the previous section you found out that fish species have their own spectral signatures and these can be used for identification purposes on satellite images.

Some fish species are known to concentrate at the junction of two water masses which are at different temperatures. Such information can be detected with infra-red sensors and can be used in fishing and in the assessment of fish stocks.

These recently developed remote sensing techniques involving satellite imagery will no doubt continue to be improved by the development of new sensing devices and an improvement in resolution. Continued efforts to find correlations between satellite images and ground studies will reap benefits.

At the World Food Conference in Rome in 1975, it was recommended that there should be a global information and early warning system with respect to food and agriculture. Management of these crucial resources could depend to a large extent on these advanced remote sensing techniques.

Active sensing

In this type of sensing the energy waves are produced by a special device and then picked up by appropriate sensors.

In *one-way active sensing* miniature high-frequency transmitters are implanted in an animal or attached to a collar around its neck. Each animal can be given a particular wavelength so that its movements can be tracked using directional radio-receiving equipment. This technique has proved to be very successful in tracking and mapping the territories of nocturnal animals such as feral cats, urban foxes, badgers and squirrels, and animals living in

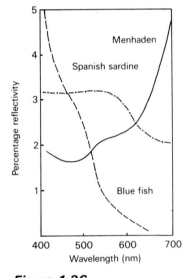

Figure 1.36
Spectral signatures of three fish species

**Figure 1.37
Transmitter attached to
the collar of a bear**

isolated habitats like the moose in Alaska, or the grizzly bear in the United
States of America.

Two-way active sensing occurs when the image is formed by the return from
the object of energy waves that have been produced by the system. Examples
are ***radar*** (**RA**dio **D**irection **A**nd **R**anging) which uses microwaves, and ***echo
sounding*** and ***sonar*** (**SO**und **NA**vigation **R**anging) which use sound waves.

The main advantage of radar is that it can penetrate cloud and fog and so it is
often the only method of remote sensing that can be used when imaging parts
of the world like tropical rain forests, which are covered in cloud for most of
the year. Most radar systems scan at an angle from the aircraft, so that relief
features are exaggerated. Crops, vegetation patterns and even certain insects
like locusts can be distinguished.

Sound waves, unlike electromagnetic waves, travel quickly and well under
water. They are useful for remote sensing in devices like echo sounders and
sonar, which emit series of short sound pulses concentrated into a narrow
beam. Both of these systems are widely used in the fishing industry. Echo
sounding involves sending sound waves from a vessel vertically downwards
to the seabed and then picking up the 'echo'. Sonar emits the sound waves in
horizontal directions. The received sound waves are converted into electrical
impulses which are amplified and displayed on an oscilloscope or chart
recorder.

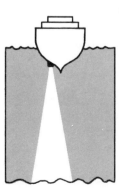

**Figure 1.38
Echo sounding**

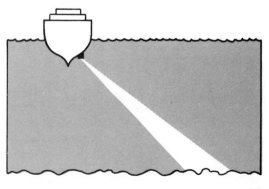

**Figure 1.39
Sonar**

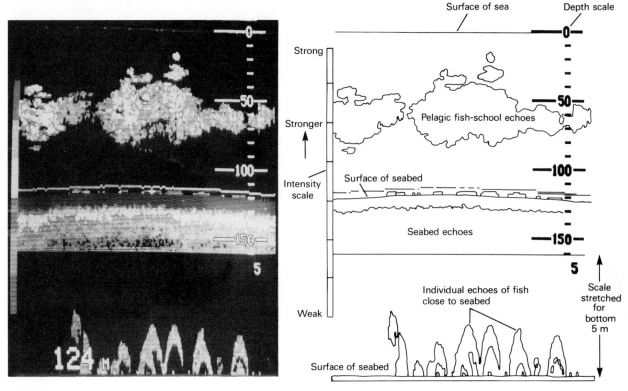

a Echogram

b Interpretation of echogram

Figure 1.40

Echograms can provide information about the daily behaviour-patterns of fish, and their behaviour during fishing. These data can then be used to predict the occurrence of fish in certain positions, and influence the methods of catching them.

1.1.8 Three dimensions from two dimensions!

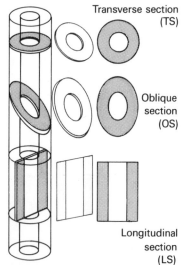

Figure 1.41
LS, TS and OS cut from a hollow cylinder

Most observations in biology are made using two-dimensional microscope sections and photographs. There are three types of section:

(i) A *longitudinal* section (**LS**) is taken through the long axis of the specimen.

(ii) A *transverse* section (**TS**) is taken through the short axis of the specimen, that is, at right angles to the LS.

(iii) An *oblique* section (**OS**) is taken through the specimen at any other angle.

Many students have great difficulty in relating two-dimensional images in the form of sections, drawings or photographs, with the three-dimensional appearance of the complete specimen. Particular difficulties arise if the specimens produce the same image in TS and LS. Only when both the TS and LS images of the same specimen are considered together can the most probable three-dimensional shape be predicted.

Figure 1.42
TS, LS and 3-D shapes

TS image	LS image	Probable 3-D shape
		Sphere
		Rod or cylinder
		Ellipse or cigar

1 Copy Figure 1.43 and for each of the sections shown, predict the most probable appearance of the missing section or three-dimensional shape either by drawing or explaining its shape.

Figure 1.43
2-D versus 3-D

TS image	LS image	Probable 3-D shape
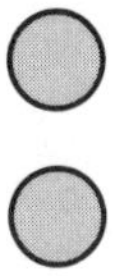	(i)	Ring-like or 'polo' shape
	 (ii)	
(iii)		Two tubes lying alongside each other. Composed of spherical subunits
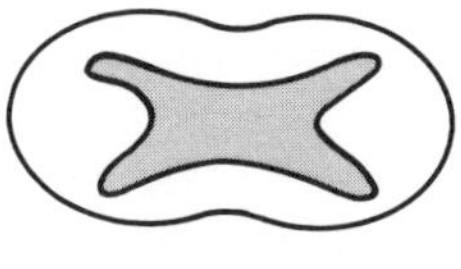	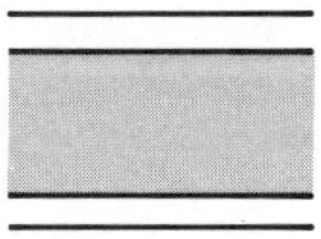 (iv)	

2 Figure 1.44 shows part of a secretory cell in the tegument of a parasite. Predict the most likely three-dimensional appearance of these secretory vesicles.

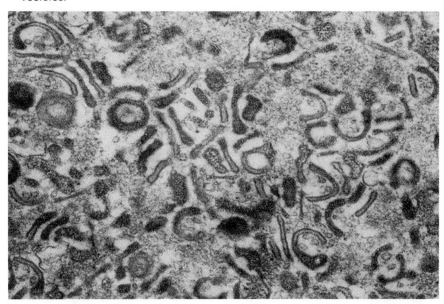

Figure 1.44
Secretory vesicles from a trematode integument cell

3 Figure 1.45 shows a TEM containing an LS of a protoplast, an immature chloroplast in a plant cell. What are the possible three-dimensional shapes of
(i) the protoplast itself?
(ii) the internal membranes highlighted in the box?
Use only the visual information provided when making your prediction.

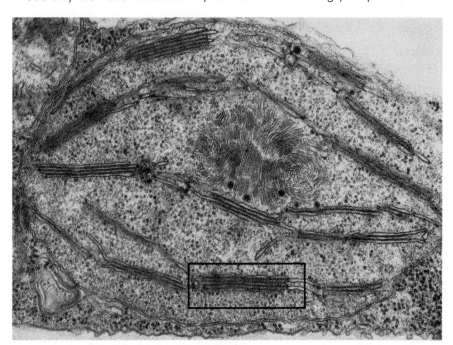

Figure 1.45
TEM protoplast

1.1.9 Magnification and scales

A vital piece of information which is often forgotten when recording observations is an indication of the size of the image recorded. This may be done in several ways:

(i)   By including in the recording an image of a familiar object, for example a pen, a penny, etc. at the same size or scale. This can at best give only a relative sense of size or proportion.

(ii) By including in the recording an image which shows the specimen at its normal size – see Figures 2.53 and 2.54 of *Drosophila*, page 68. Again this gives a relative sense of size and proportion but it is also possible to determine an approximate magnification for the recording. This technique is limited to specimens which fall well within the resolution of the unaided eye.

(iii) By including in the recording a scale which is at the same magnification as the specimen – see Plates 2.4(a) and (b), page 63. Using this technique, which can be applied to the full range of biological observations, the actual size of the specimen or part of the specimen may easily be determined by measurement.

(iv) By including, along with the title of the recording, the magnification at which the recording was made, usually by showing the number of times the actual specimen has been magnified or reduced. This is conventionally given as a multiplication sign ($\times$) followed by the numerical value – see Plates 2.3(a) and (b), page 60.

Since magnification in biology is normally taken as the ratio of the linear measurements of the recorded image to the actual size of the specimen, converting from scale lines to magnification and vice versa is a simple task:

$$\text{magnification} = \frac{\text{length of scale line given}}{\text{length of scale line measured}}$$

(both lengths must be converted to the same unit)

1 Using the information given above, determine the magnifications of Plates 2.4(a) and (b), page 63.

2 For Plates 2.3(a) and (b), page 60, calculate the lengths of scale line you would have to draw to represent 1 μm and 100 μm respectively.

3 Similar calculations may be practised on many of the figures and plates given in this book.

1.2 RECORDING

1.2.1 Drawings and diagrams

Biology involves a considerable amount of time examining structures, their relationships to each other and their relationships with their functions. For this reason, the accurate drawing of biological objects is a skill which has always been central to biological communication. Although drawing has been superseded in institutions of research and higher education by photography, drawing still holds an important place in schools and colleges.

All Advanced-level syllabuses list drawings and diagrams as important means of communicating biological information, which will be assessed in the examination. However, training in this particular skill is usually haphazard, students being expected to acquire the skill by just being exposed to it. This skill requires specific instruction and practice if it is to be a useful form of communication. Section 2 is concerned solely with the skill of drawing. By following the examples carefully and then completing the exercises you should become competent at biological drawing.

1.2.2 Tables

We have already seen that modern biology requires observations to be quantified wherever possible. By quantifying we mean that the observation is expressed as a number, which involves counting or measuring. When we count things, we are quantifying natural or experimental observations of a discrete nature, whether they are the number of dark bands on the shell of a snail, the number of bubbles produced per minute by a piece of *Elodea* or the number of vestigial-winged *Drosophila* in a population.

When observations are measured the data are continuous: involving length, weight, volume, time, etc.

All raw data are best recorded in tabular form, the type of the table depending on the nature of the variables being observed. It is always better to construct a suitable table prior to data collection so that the table can be completed as you count or measure; this improves accuracy and saves time. There are many different types of table, particular types being required for particular types of data, for example, independent or dependent variables; interdependent variables; frequency distributions, etc. It is important to know the different types of table and when they should be used. Various situations and the appropriate tables are described in Volume 1 of the series, *Dealing with Data*, Section 2, pages 6–11. You will also get plenty of practice at table design as you complete the examples and exercises in later sections.

1.2.3 Graphs

Once the data have been recorded in a table they can be drawn as a graph. The beauty of graphs is not only that a lot of information can be presented within the visual field, but that the data can become more meaningful. In certain instances it is better to record the data directly on to a pre-prepared graph. There are many different types of graph and it is very important to construct the right one for the particular type of data under consideration. You should be aware of the different types of grids and graphs that are available. A wide range of graphs are illustrated in *Dealing with Data*, pages 11–49.

1.3 INTERPRETING

1.3.1 Drawings and diagrams

The accurate representation of biological structures with drawings requires that the structures are recognised and labelled. The correct relationships between the structures, and between the structures and the functions they perform must be understood – this can be appended to the drawings as annotations. As you work through Section 2 you will be given plenty of opportunity to develop these skills.

1.3.2 Graphs

If quantitative data are displayed in an appropriate graphical form, it is often possible to identify trends and relationships. In situations where the variables are independent or dependent, it is possible to devise equations relating the variables, if the graph approximates to a straight line – see *Dealing with Data* pages 50–6. When the variables are interdependent, a scattergraph can show the intensity of the relationship, and a line of best fit can be used to determine the relationship between the variables – see *Dealing with Data* pages 56–9. Finally, with frequency distributions, it is possible to determine if the data forms a 'normal' distribution – see *Dealing with Data* pages 61–3.

The potential of graphs in the initial interpretation of data should not be underestimated – they not only give a good visual impression of what is happening, but can also assist in interpretation.

1.3.3 Statistics

Statistics are extremely useful in biology since they can be used to make sense out of seemingly confusing data. Because biological data are so variable it is often difficult, if not impossible, to come to any firm conclusions about them. The use of statistics helps us to come to some tentative conclusions within specified probabilities. Most of the time only three statistical tests of significance are required:

(i)   When the data contain samples from which means, and standard deviations can be calculated, the t test is used – this can involve counts or measures.

(ii) If the data consist of purely discrete data then the Chi^2 (χ^2) test is used. It is particularly useful when the data have been arranged in a contingency table.

(iii) When the data are interdependent then the correlation coefficient is used.

These three tests are explained and illustrated in detail in *Dealing with Data* pages 87–112.

Plenty of opportunities are given in the examples and exercises for you to use graphs and statistics to analyse data.

2 Drawing

The teaching and development of this skill has normally been left to chance in the past. Students who didn't already have the ability to draw were not taught how to do so. The importance of developing this skill cannot be overemphasised.

In Section 1.1.4, page 7, the similarities between biology and engineering were explained – both essentially deal with the relationships between structures and their functions. Both disciplines use a particular form of drawing – the engineers call it an engineering drawing or a mechanical drawing. Their line drawings specifically focus attention on structures and their inter-relationships. Biology also requires a special kind of drawing which we shall call a *biological drawing*. Both engineering and biology have to contend with the problem of representing three-dimensional structures. For this, drawing can be clearer and more descriptive than photographs. Both types of drawing are normally labelled and have appended explanatory notes or annotations.

2.1 THE PRINCIPLES OF DRAWING

An important part of the study of biology is gaining knowledge and evidence of the architecture of living things. Drawings act as a pictorial record of observations and interpretations, enabling us to record patterns and characteristics.

Using drawings it is easy to describe the complex relationships of parts of organisms and highlight structures of biological significance.

Examiners therefore find that they can readily assess a candidate's ability to recognise biological structures and their inter-relationships by including drawing as one of the assessed skills. The Advanced-level syllabuses list among the assessment objectives, 'to communicate clearly, and interpret, biological information by means of diagrams and drawings'. This includes drawing accurately to scale, showing the relationship between the form of a structure and its function, and revealing accurate observation.

Syllabuses usually require something in between a drawing and a diagram. Examiners sometimes refer to this type of drawing as a tissue plan, block diagram or simplified sketch; we have suggested *biological drawing*. This type of illustration can only be produced after some specific biological training.

2.1.1 Biological drawings: the requirements

What is the difference between a drawing and a diagram? A *drawing* is a detailed and accurate representation of a specimen. No biological knowledge is needed for this, and an artist could probably do this best. A *diagram* is accurate in its general proportions, but is very stylised and only shows the

a Photograph

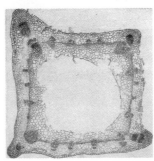

b Drawing

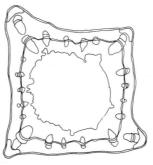

c Biological drawing

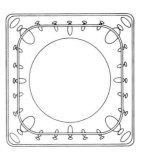

d Diagram

Figure 2.1
TS stem of Vicia faba

important regions of the specimen. Textbooks usually show diagrams rather than drawings.

Examine the illustrations on this page. They show photographs, drawings, biological drawings and diagrams of a transverse section of a plant stem and a TEM of a plasma cell.

Examine Figures 2.1 and 2.2 carefully and note the differences and similarities between each of the illustrations.

a Photograph

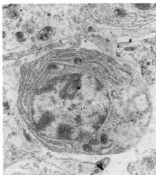

b Drawing

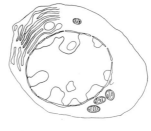

c Biological drawing

d Diagram

Figure 2.2
TEM plasma cell

1 What are the differences between a biological drawing and a drawing?

2 What are the differences between a biological drawing and a diagram?

You will be given plenty of opportunities to produce biological drawings during this course. These are what you may be required to produce in an examination, though they may be called by an alternative name.

It is important when making a biological drawing that the following points are carefully noted:

(i) Make sure that the relationships and proportions of the parts are correct. You may have to enlarge what is being observed, for example when drawing from a microscope, or reduce the size if the specimen is very large.

(ii) In all cases get into the habit of drawing large diagrams. The actual size will of course depend on the size of the page and how much room has to be left for labels and notes around the diagram.

(iii) Always use a sharp HB pencil, which produces clear lines which are easy to erase if mistakes are made. Draw firm continuous lines, not scratchy lines (see Figure 2.3).

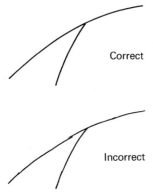

Correct

Incorrect

Figure 2.3
Correct and incorrect lines

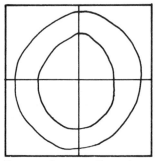

Four equal squares if specimen
is radially symmetrical

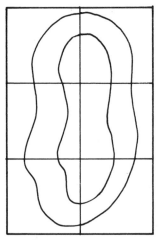

Six equal squares if specimen
is an elongated shape

**Figure 2.4
The divided box**

A method which helps to ensure that the proportions are correct is to draw a box of a suitable size and divide it into equal squares, four if the specimen is radially symmetrical, or six squares if it is elongated.

Take each square in turn and draw light marks as guides to accurate shapes and proportions. Then draw in the outlines and finally complete the details.

Step 2: Draw proper lines
for inner and outer
boundaries

Step 3: Draw the
epidermis and cambium

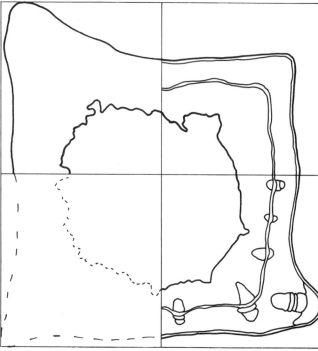

**Figure 2.5
Stages of
outlining and
filling in**

Step 1: Draw light marks
to give an idea of inner
and outer boundaries

Step 4: Complete by
filling in the remaining
structures

**Figure 2.6
Drawing from a
microscope**

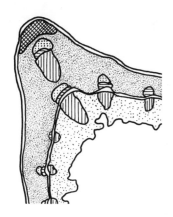

Figure 2.7
Using cross-hatching and dots

When drawing from the microscope, the paper should be positioned on the right-hand side of the microscope (for right-handed people), while the left eye observes the specimen. By keeping both eyes open one can learn to draw while looking at the page with the right eye and at the same time continually observing the specimen with the other. It is important to learn the skill of **relaxed viewing**. Often students complain about eye strain and headaches after a prolonged session of looking down a microscope. This discomfort can be eliminated by keeping both eyes open when looking down a monocular microscope. At the beginning, you could cover the eye not being used with your hand, and then gradually take it away. After a while your brain eliminates what the unnecessary eye sees.

The differences between regions of a biological drawing are best shown by a system of cross-hatching or by an increasing density of dots. Do not try to shade in areas as this is rarely successful and the pencil tends to rub off on to the adjacent page.

2.1.2 Labelling

Labelling a biological drawing ensures that you are able to recognise specific structures and helps you to remember what they look like. Usually the specimen or photograph of an actual specimen used in an examination is new to the candidates and so they have to relate their novel observations to what they have learned and experienced – labelling thus involves interpretation. Careful and accurate labelling is just as important as the actual drawing, and should be done neatly and clearly. The labels should be printed in capital letters and should not be written on the drawing, but arranged well away from it. The labels should be connected to the drawing by leader lines. These lines should be parallel wherever possible to the top edge of the page and should not cross one another so that there is no confusion regarding what structure the label refers to. The point to which the label refers should be shown precisely by means of an arrowhead or a large dot at the end of the leader line.

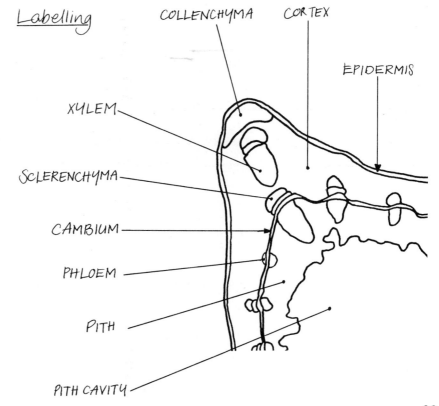

Figure 2.8
Labelling and leader lines

2.1.3 Annotating

Annotating means that you include short explanatory notes in brackets below the labels on the biological drawing. In an examination, these should only be included when requested to do so. Otherwise you should always annotate whenever you draw and label. Annotating helps you to highlight the biological significance of the structures, especially regarding their functions.

Finally, as well as a full title, a useful addition to a biological drawing is the approximate magnification, for example ×1, ×5, × ½ and so on.

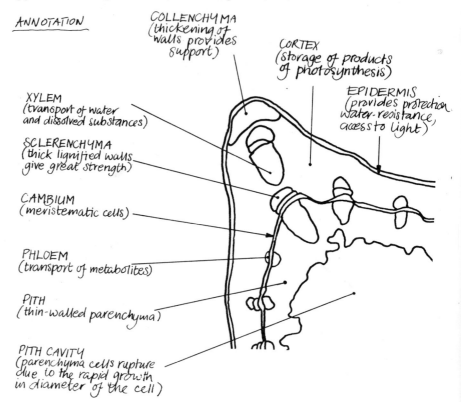

Figure 2.9
Annotation of labels

2.2 LEARNING HOW TO DRAW

In biology, specimens can range from large whole plants or animals to very tiny microscopic structures. In recent years the transmission and scanning electron microscopes have been used to greater and greater effect. Each separate situation requires a different approach to drawing, and so the best way of acquiring this skill is to find out how it is applied in a variety of different situations.

The unit of life is the cell. Unicellular organisms are able to survive, grow and reproduce. Multicellular organisms have evolved a system of division of labour whereby cells become **differentiated** so that they carry out particular functions. All cells have within them a number of different **organelles** which carry out specialised functions. As you saw in Section 1, electron microscopy has greatly enhanced our knowledge of these organelles. Groups of similar cells are often closely associated with each other in *tissues*; tissues are organised in particular ways to form *organs*, which in turn are linked together in particular ways to give the *systems* of which the *organism* is composed. An understanding of such hierarchies is an important aspect of biology.

2.2.1 The cell

The basic unit of both structure and function in living organisms is the *cell*. The study of cells, particularly using microscopes, is called *cytology*. This has advanced dramatically over the past few decades, due largely to the development and application of the electron microscope. Because of the high magnifications and good resolution of the TEM, various structures within cells have been discovered, and existing structures found to be much more complex than when observed under the light microscope. These structures are in the main new to you and so first of all you will have to become familiar with them and be able to recognise them in different cells. You will also have to understand what the functions of these organelles are. The following section deals with each of these organelles in turn, using photographs and annotated biological drawings.

The cell membrane (plasma membrane)

This cannot be seen with the light microscope but can be observed in electron micrographs as a thin double line, indicating that it has three layers. It consists of a molecular sandwich of a lipid bilayer between two protein layers, and acts as a differentially permeable membrane. Water and solutes pass through by diffusion, osmosis and active transport; suspended materials can get into the cell by *phagocytosis* and tiny droplets by *pinocytosis*.

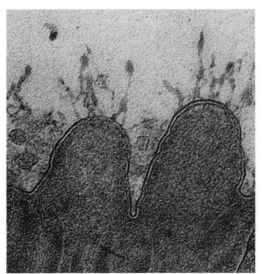

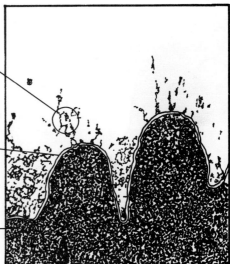

Long chain carbohydrates attached to proteins in the membrane to form a glycocalyx

Plasma or cell membrane – electron-opaque lipids arranged in two layers. Appears as two dark lines approximately 7.5 nm apart

Electron-opaque particles give the cytoplasm a medium density

a TEM cell membrane (×130 000)   *b Drawing*

Figure 2.10

The cell wall

Most plant cells are bounded on the outside by a thick and rigid wall, which consists of microfibrils of cellulose embedded in a matrix of hemicelluloses and pectic substances. Between the cellulose of neighbouring cell walls is the *middle lamella* which consists of calcium and magnesium pectates and noncellulose polysaccharides. This helps to hold the cells in place. Neighbouring cells are connected by **plasmodesmata**, which are fine pores in the cell walls.

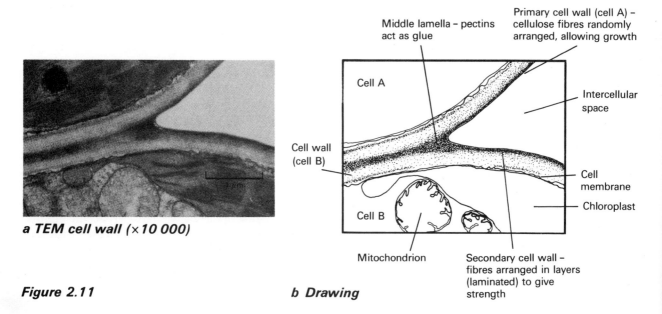

a TEM cell wall (×10 000)

Figure 2.11

b Drawing

Labels:
- Middle lamella – pectins act as glue
- Primary cell wall (cell A) – cellulose fibres randomly arranged, allowing growth
- Cell A
- Intercellular space
- Cell wall (cell B)
- Cell membrane
- Cell B
- Chloroplast
- Mitochondrion
- Secondary cell wall – fibres arranged in layers (laminated) to give strength

Microvilli

Some animal cell membranes have tiny projections called **microvilli** which increase the surface area and thus aid absorption. These microvilli can just be seen with the light microscope as a fuzzy line and hence are known as the **brush border**.

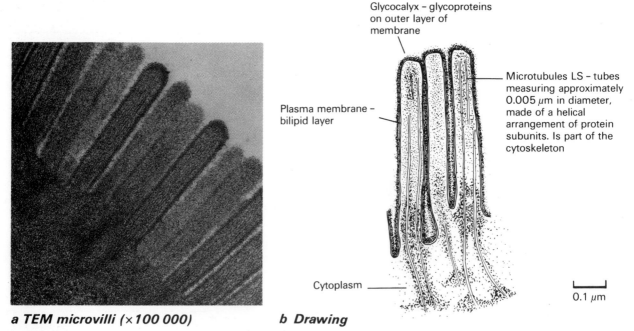

a TEM microvilli (×100 000)

b Drawing

Figure 2.12

Labels:
- Glycocalyx – glycoproteins on outer layer of membrane
- Plasma membrane – bilipid layer
- Microtubules LS – tubes measuring approximately 0.005 μm in diameter, made of a helical arrangement of protein subunits. Is part of the cytoskeleton
- Cytoplasm
- 0.1 μm

Endoplasmic reticulum (ER)

The endoplasmic reticulum is a double membrane extending throughout most of the cytoplasm and was discovered by electron microscopy. It consists of an extensive three-dimensional network of canals and sacs (**cisternae**), connecting the plasma membrane with the nuclear membrane. The ER is involved in the synthesis and transport of substances.

Figure 2.13

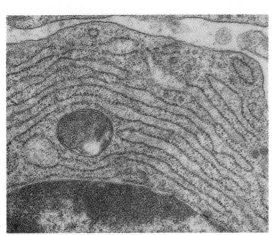

a Rough ER (×23 500)

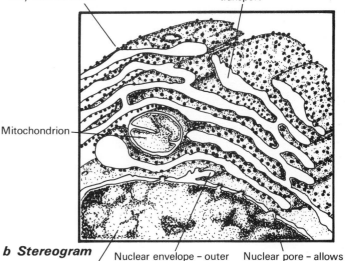

Ribosomes on outer surface of ER – protein synthesis lipids may also be found

Membranous sacs of ER cisternae – interconnections allow transport

Mitochondrion

b Stereogram

Nucleus

Nuclear envelope – outer membrane connects with ER

Nuclear pore – allows mRNA into cytoplasm

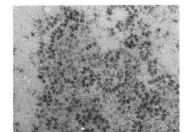

**Figure 2.14
Ribosomes (×40 000)**

Figure 2.15

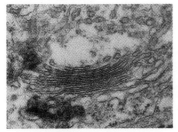

a Golgi apparatus (×28 000)

There are two forms:

(i) **rough ER** in which the membranes are covered on the cytoplasmic side by ribosomes (see below). This is associated with protein synthesis.

(ii) **smooth ER** in which there are no ribosomes present. This is associated with steroid synthesis.

Ribosomes

These are very small structures, only about 15–20 nm in diameter, and are found in large numbers throughout the cytoplasm in both prokaryotic and eukaryotic cells (see pages 49 and 50). As you have seen, many are associated with the rough ER. When extracted from cells they often form clusters or **polyribosomes** (**polysomes**). Ribosomes are involved in the synthesis of proteins. During this process amino acids are linked together one by one to form long polypeptide chains. The ribosome acts as a binding site where the appropriate amino acid is added to the chain as it is synthesised.

The Golgi body (Golgi apparatus)

Named after its discoverer, the Golgi apparatus is found in nearly all eukaryotic cells. It resembles smooth ER and is connected with it both structurally and functionally. It consists of flattened cisternae (compartments) plus an associated system of **Golgi vesicles**. In plant cells separate stacks called **dictyosomes** can be distinguished. The Golgi apparatus is involved in combining carbohydrates with proteins (**glycosylation**) to form glycoproteins for export from the cell.

Membrane of cisterna   Lumen of cisterna

Peripheral tubules

Secretory vesicles

Dictyosome (stack of cisternae)

Ribosomes

Rough ER

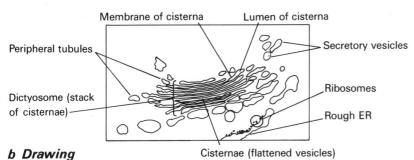

b Drawing

Cisternae (flattened vesicles)

c Stereogram

Lysosomes

These are small (0.2–0.5 μm) sacs containing digestive enzymes, and have a single membrane. They are produced by the Golgi apparatus and so are found in those cells that possess one. They can digest practically any cellular material – this is why they are separated from the rest of the cytoplasm by membranes. Lysosomal enzymes are obviously a potential hazard to a normal cell, and its survival depends on the ability to maintain the membranes of the lysosome. Various substances, including silica, are known to increase the fragility of the lysosomal membranes and so are capable of causing destruction of cells.

Mitochondria

These are small organelles that can vary considerably in shape and size (1–10 μm long) and they have a complex internal structure. They are found only in aerobically respiring eukaryotic cells. The number present in the cytoplasm can vary from about 50 to 2500 depending on the type of organism and nature of the cell. The greater the number, the greater the metabolic activity of the cell. Mitochondria have their own DNA which is ring-shaped and attached to the boundary membrane. These organelles are bounded by a double membrane, the inner of which is infolded to form **cristae** which extend into the interior, thus increasing the surface area. The interior of the mitochondrion contains a semi-ringed matrix. Enzymes of the tricarboxylic cycle are found inside the mitochondrion, both in the matrix and associated with the membranes. Particles on the matrix side of the cristae appear to be associated with the electron transport system in the production of ATP (oxidative phosphorylation).

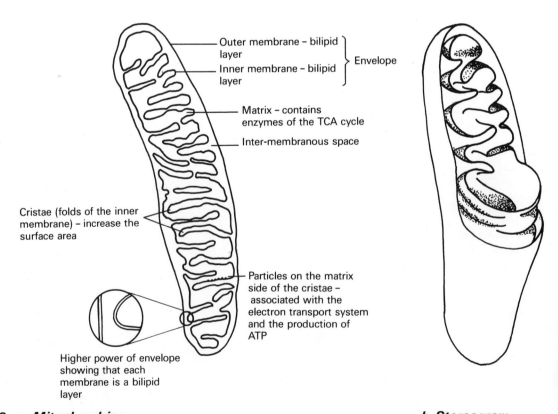

Outer membrane – bilipid layer ⎫
Inner membrane – bilipid layer ⎬ Envelope

Matrix – contains enzymes of the TCA cycle

Inter-membranous space

Cristae (folds of the inner membrane) – increase the surface area

Particles on the matrix side of the cristae – associated with the electron transport system and the production of ATP

Higher power of envelope showing that each membrane is a bilipid layer

Figure 2.16 a Mitochondrion   *b Stereogram*

Plastids

These are double membrane organelles found only in the cytoplasm of plant cells. There are several different types:

(i) **Chloroplasts** are lens-shaped and vary in number from 1 to about 100 per cell. They are about 3–10 μm in diameter and thus are visible with the light microscope. They contain chlorophyll and other photosynthetic pigments and are the site of photosynthesis.

Thin TEM sections show the outer membrane to be smooth while the inner one extends inwards to form a system of layers called **lamellae** where the pigments are located. These lamellae often appear as flat discs called **thylakoids** which are stacked together to form **grana** and they are interconnected by **intergrana lamellae**. This membrane system is the site of the **light reactions** of photosynthesis, containing the necessary pigments, enzymes and electron acceptors. These membranes lie in a matrix called the **stroma** which is the site of the **dark reactions** of photosynthesis, containing enzymes involved in the Calvin cycle together with sugars and organic acids. The products of photosynthesis in the form of starch grains and lipid globules are often found in the stroma.

(ii) **Chromoplasts** contain carotenoid pigments and are mainly found in fruits and flowers. They cause the bright colours.

(iii) **Leucoplasts** are colourless and are found mainly in food storage organs.

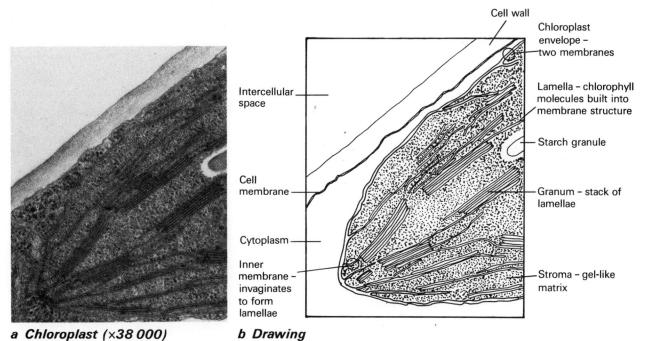

a Chloroplast (×38 000)   *b Drawing*

Figure 2.17

Microtubules and microfilaments

Microtubules are, as their name implies, very fine tubes found in the cytoplasm of eukaryotic cells. They are found in a number of different forms which are related to their function.

(i) **The cytoskeleton:** since microtubules are long and fairly rigid, they help to determine and maintain the shape of cells.

(ii) **Centrioles:** animal cells and a few primitive algal cells contain two small hollow cylinders known as centrioles. They are found near the nucleus in a region called the **centrosome**. They lie at right angles to

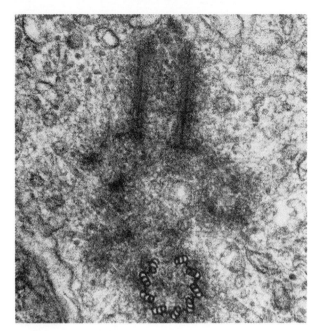

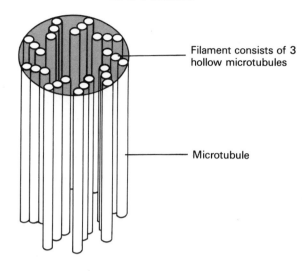

Circle of 9 filaments

Filament consists of 3 hollow microtubules

Microtubule

a Centriole TS and LS (×39 000)

Figure 2.18

b Arrangement of microtubules in the centriole

one another just outside the nuclear envelope. Each centriole has a very regular cross-sectional pattern of nine triplets of microtubules. The function of the centrioles is to organise the complex array of microtubules known as the *spindle apparatus*, which is utilised in cell division to separate the chromosomes. At the start of cell division, the centrioles replicate and the two pairs move to the poles of the cell, the spindle radiating from them towards the equator.

(iii) *Cilia and flagellae:* these are fine hair-like structures found in many types of eukaryotic cells which extend outwards from the plasma

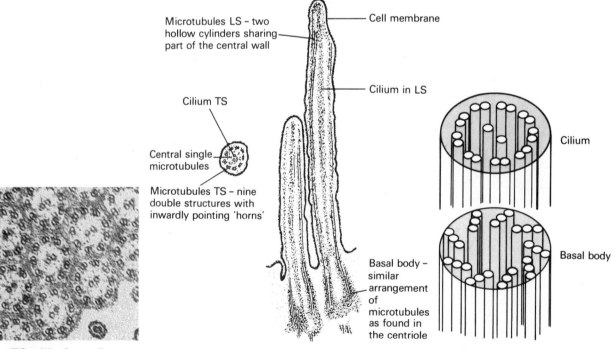

Microtubules LS – two hollow cylinders sharing part of the central wall

Cell membrane

Cilium in LS

Cilium TS

Central single microtubules

Microtubules TS – nine double structures with inwardly pointing 'horns'

Basal body – similar arrangement of microtubules as found in the centriole

Cilium

Basal body

a TS cilia from flame cell (×40 000)

Figure 2.19

b Cilium TS and LS

c Arrangement of micro-tubules in cilium and basal body

membrane. They are involved in movement either of the cells themselves or of substances past the cells. Cilia are much shorter than flagellae but have the same basic structure – nine pairs of microtubules forming a ring surrounding two microtubules in the centre.

Interestingly, both cilia and flagellae arise from **basal bodies** which have the same microtubule structure as the centrioles.

Nucleus

The nucleus is the largest organelle (10–20 μm long) in the cytoplasm of eukaryotes and so can easily be seen with the light microscope. It is usually spherical and bounded by a double membrane known as the **nuclear membrane** or **nuclear envelope**. The outer layer is often covered with ribosomes and is contiguous with the ER. The space between the nuclear membranes is called the **perinuclear space**. The nuclear envelope is richly perforated with **nuclear pores** which are readily seen in freeze-etched electron micrographs (see page 20). These pores are covered by a thin single membrane and allow the selective exchange of substances between the nucleus and the cytoplasm.

The **chromosomes**, consisting of DNA and proteins called histones, are found in the nucleus, but these are only visible when the cell is dividing. When the cell is not dividing, the chromatic material exists only as fine threads of **chromatin**. At this time (interphase), we can distinguish between dense darkly stained patches of **heterochromatin** which occur just inside the nuclear envelope, and more diffuse **euchromatin** found nearer the centre of the nucleus. RNA is also present. The nucleus not only carries the hereditary information in the chromosomes, but it also controls the activities of the cell.

The most noticeable body within the nucleus is the **nucleolus** which consists mainly of nucleic acids and protein. It is where **ribosomal RNA** is manufactured.

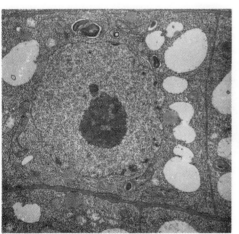

Cytoplasm

Cell wall

Nuclear envelope – double membrane

Vacuole – apparently empty, but contains cell sap

Nuclear pore – allows RNA to move from nucleus to cytoplasm

Tonoplast – single membrane

Chromatin – complex of DNA and histone proteins

Nucleolus – site of RNA production

a Nucleus and vacuoles (×8 400)

b Drawing

Figure 2.20

Vacuoles

Most plant cells have a large fluid-filled cavity in the cytoplasm called a **vacuole**, which is bounded by a unit membrane called the **tonoplast**. The vacuole can often occupy as much as 80–90 per cent of the total cell volume. The fluid contents of the vacuole consist mainly of water and solutes, although it is sometimes called the **cell sap** since sucrose is often stored there. Vacuoles have an important role in maintaining the turgidity of plant cells.

Prokaryotes and eukaryotes

Since the development of the electron microscope, it has become evident that the living world is made up of two quite different types of cell. Although both these types are very similar at both the molecular and the functional levels, they show many structural differences.

Many taxonomists now use these two cells types – the **prokaryotes** and the **eukaryotes** – as the basic division of all life into two super-kingdoms.

Plate 2.1 contains two drawings of prokaryotes: (a) a blue-green alga and (b) a bacterium.

Plate 2.2 contains two drawings of eukaryotes: (a) a plant cell and (b) an animal cell.

1 Referring to suitable textbooks and to the information given previously in this section about the various cell organelles, label copies of the four diagrams.

2 Copy out the following table, and using the information both from the diagrams and which you have gleaned from other sources, complete the table and the questions.

Characteristic	Prokaryote cell	Eukaryote cell
Size (μm)		
Cell wall		
Plasma membrane		
Form of DNA		
Position of DNA		
Membrane-bound inclusions		
Structure of cilia		

3 Which of the two types of cell (prokaryote or eukaryote) is more complex?

4 One theory concerning the evolution of these two cell types suggests that eukaryotes arose by the enclosure of one prokaryote cell within another's cytoplasm. Which of the membrane-bound inclusions of the eukaryote cell resembles, in size and form, a prokaryote cell?

5 Make out a table with the following headings. Complete the table by entering the structures and organelles. Put a tick in the appropriate box if present, and a cross if absent.

Structure	Plant cell	Animal cell

etc.

PLATE 2.1

a

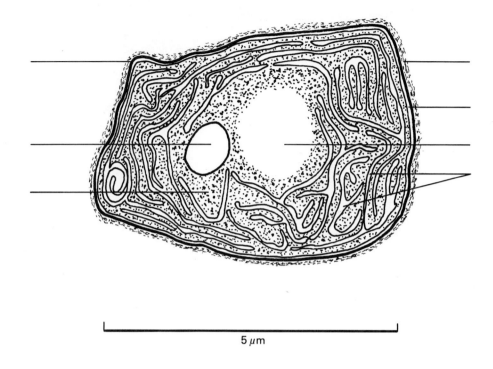

5 μm

b

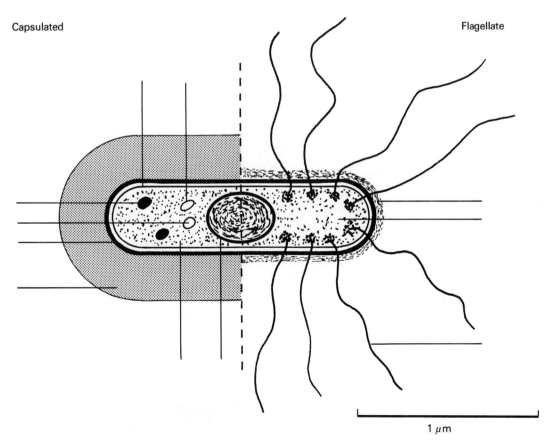

Capsulated   Flagellate

1 μm

PLATE 2.2

a

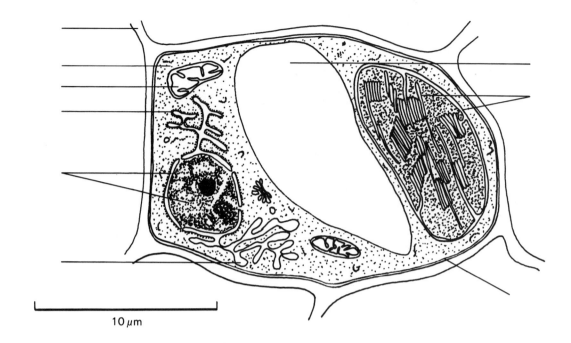

10 μm

b

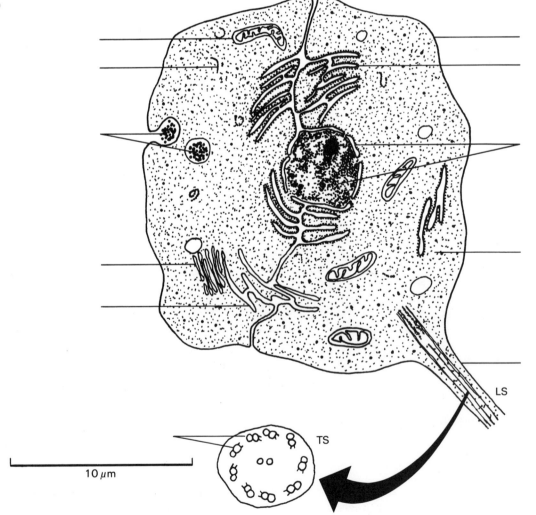

LS

TS

10 μm

2.2.2 Angiosperm histology

The division of labour within each flowering plant has resulted in particular cells specialising in one process within the organism. Such specialisation, termed **differentiation**, leads to a characteristic structure.

As you found out on pages 40–1, a tissue can be defined as a group of specialised cells lying close together and carrying out the same function. This section is concerned with the structures and functions of the most important plant tissues.

Meristematic tissues

A tissue in which a high proportion of the cells are dividing, the main function being the production of new cells, is called a **meristem**. Characteristic of all flowering plants are the **apical meristems**, which occur just behind the tip of each shoot and root. New cells produced by these meristems add to the height of the shoot or the length of the root and in so doing develop into the **primary structures** of the plant. Other meristems occur elsewhere in the plant, for example in parts of the stem and the leaf sheath, and also add to the primary structures. Such meristems have differentiated tissues both above and below them and are described as **intercalary meristems**.

Some plants, notably the trees, show considerable increase in girth during their growth. This is due to another meristem, the **cambium**, which because of its position and the way it adds new cells to the plant, is described as a **lateral meristem**. These meristems add width, strength and increased transport capacity to the plant. The tissues which they produce are sometimes called the **secondary structures**.

Epidermal tissues

The outermost layer of cells in the younger parts of the plant make up what is called the **epidermis**. As the plant grows older, the epidermis may be replaced by other tissues, as in the formation of bark by the trunk of the tree, but it is permanent in such organs as the leaves. Because of its position, the epidermis possesses a number of unique properties.

1. It is a protective layer, being the first line of defence against parasites, predators, abrasion and the weather.
2. It is strong in its aerial parts, but this is combined with suppleness and elasticity, allowing flexibility and growth.
3. It is transparent, allowing the entry of light for photosynthesis.
4. It allows exchange of gases with the atmosphere yet prevents desiccation.
5. It is strong in its root epidermal tissues for protection, while remaining permeable to water and minerals.

The aerial parts, with their more exacting requirements, have a more complex epidermal structure. A leaf or a young stem generally has an epidermis consisting of an impermeable layer perforated at intervals by **stomata**, which provide pores of variable aperture. Cells of the impermeable layer usually appear square or rectangular when seen in TS, but when seen in surface view each cell often has a wavy outline rather like a piece of a jigsaw puzzle. Epidermal cells shaped like this are formed in many dicotyledons. The cells fit closely and exactly together, giving coherence to the layer of cells and an absence of intercellular spaces. The epidermal cells of most monocotyledons are rectangular in surface view, with the long axis of each cell parallel to the long axis of the leaf. These also fit closely together, and there are no intercellular spaces.

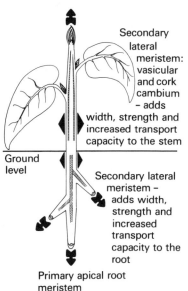

Primary apical shoot meristem – increases overall height of plant

Secondary lateral meristem: vasicular and cork cambium – adds width, strength and increased transport capacity to the stem

Ground level

Secondary lateral meristem – adds width, strength and increased transport capacity to the root

Primary apical root meristem

Figure 2.21 Meristems

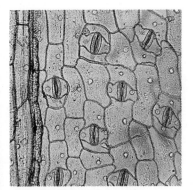

Figure 2.22 Dicotyledon leaf epidermis (×40)

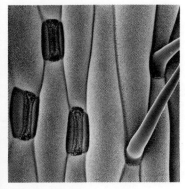

Figure 2.23 Monocotyledon leaf epidermis (×310)

The outer walls of the epidermal cells are impermeable to water, due to the addition of a fatty substance called **cutin**. This is produced by protoplasts in the cells and deposited in the cellulose wall where it moves outwards and accumulates on the surface. The layers of cutin of neighbouring cells fuse to form a continuous waterproof sheet, the **cuticle**.

Plants which face a serious and frequent risk of drying out, for example those growing in hot dry places, often possess a very thick cuticle which may be supplemented by a layer of wax. At the other extreme, plants growing in damp, shady places have a very thin cuticle.

A common feature on the epidermis are outgrowths, collectively known as **trichomes**. These show great diversity of form, from simple unicellular hairs to elaborate multicellular structures. The simplest type is the most common: the epidermal cells of most young roots develop into **root hairs**, which are unbranched and unicellular, and do not have a cuticle.

1 What is the function of these root hairs?

Another example of a trichome is the stinging hair of the nettle.

2 What do you think is the function of these stinging hairs?

Many trichomes are glandular, secreting substances which are by-products of metabolism.

Parenchyma
This is the least specialised tissue in the plant, with living cells containing living protoplasts. They have thin primary walls made of cellulose and pectic compounds, non-cellulosic polysaccharides and hemicelluloses.

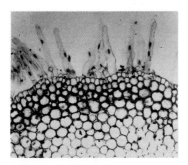

Figure 2.24
Root hairs (×50)

 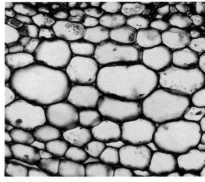

Figure 2.25
Parenchyma cells

(a) LS (×170)   **(b) TS (×130)**

Parenchyma cells are turgid, giving a certain amount of rigidity to the plant; wilting occurs when this turgor is lost. In many cases, parenchyma forms the ground substance of the plant body where it may be either closely packed or permeated by an extensive airspace system, as in aquatic stems. Parenchyma cells can change to the meristematic condition to form secondary meristems. They may also be modified in various ways to serve a variety of functions, such as photosynthesis in palisade parenchyma, aeration, food and water storage, slow conduction, absorption and protection.

3 Draw a few parenchyma cells, label and annotate.

Collenchyma
This is a living tissue made of one type of cell, which contains living protoplasts, nuclei and sometimes chloroplasts. Collenchyma cells have primary walls made of cellulose and pectic compounds. These walls may be fairly thick and are sometimes lamellated due to the alternation of layers

which are either poor in cellulose or poor in pectic compounds. The cell walls are capable of stretching, which is necessary because the cells mature and thickening begins before the elongation of that part of the plant is complete. Collenchyma provides mechanical support because of the thickening of its walls and the turgor of its cells. The cells are either short and prismatic, or long and tapering, being elongated in the same direction as the longitudinal axis of the plant. They are always polygonal in TS.

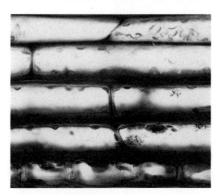

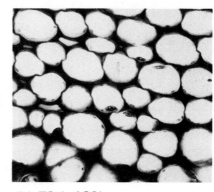

Figure 2.26
Collenchyma cells

(a) LS (×250)   *(b) TS (×160)*

Collenchyma is very common in stems and leaves where it is usually found just below the surface. It is also found in the ribs of ribbed stems and petioles. It is not found in the stems and leaves of monocotyledons which develop sclerenchyma cells early in their growth.

4 Draw a few collenchyma cells, label and annotate.

Sclerenchyma
Sclerenchyma is strengthening tissue like collenchyma, but sclerenchyma cells are normally dead when mature. This is because sclerenchyma cells have thick, non-elastic secondary walls which are impregnated with lignin. This results in the death of the living protoplast of the cell because substances cannot pass into or out of the cell. Sclerenchyma is probably the most difficult of the plant tissues to identify because its cells are often associated with functionally different parts of the plant, producing secondary structural differences in the cells.

The most common type of sclerenchyma cell is the *fibre*, groups of which appear in the xylem, phloem, pericycle and cortex (see later sections). A typical fibre is greatly elongated with its ends tapering to fine points. Such cells often lie parallel to each other in bundles, overlapping at their ends with other fibres above and below, thus forming a long strand running the whole length of the plant. The length and strength of such strands make them valuable raw materials for the textile industry, for example flax, which is used in the manufacture of linen.

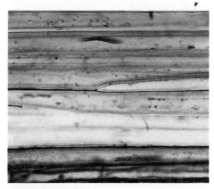

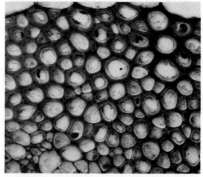

Figure 2.27
Sclerenchyma fibres

a LS (×425)   *b TS (×275)*

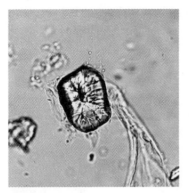

Figure 2.28
Sclereid (×160)

5 Draw a few sclerenchyma fibres, label and annotate.

Other fibres which are structurally sclerenchyma are found in the xylem, but here they are often called **xylem fibres**.

The second type of sclerenchyma cell is the **sclereid** or stone cell. These retain the essential characteristic of the tissue, a thick secondary wall, but show greater variety of shape than do fibres. Sclereids are normally not more than two or three times longer than they are broad, many being branched and contorted. Usually sclereids occur as solitary cells, but clusters and larger accumulations are also found, as in the shell of many nuts or in the flesh of a pear giving it a slightly gritty texture.

6 Draw a sclereid, label and annotate.

Xylem

Plants need to take up water and dissolved mineral salts from the soil and distribute them to all parts of the plant. Xylem vessels and tracheids make up a continuous system of fine tubes stretching from the roots to all parts of the shoots. Each section of each of these tubes is a single cell or **element**, or to be exact, the cell wall of a single cell. As the xylem differentiates, a secondary cell wall impregnated with lignin is laid down inside the primary cell wall, and in the final stages the primary wall is also impregnated. As in sclerenchyma, this results in the death of the cell. Observations of longitudinal sections of xylem vessels indicate that this secondary cell wall is not uniform. The xylem vessels which develop first within the elongation zone of the root or shoot are known as **protoxylem**. These vessels have a secondary wall capable of becoming stretched because the lignin is laid down in an annular or spiral pattern. Those xylem vessels which are formed later, the **metaxylem**, have massively thickened secondary walls, showing a reticular or pitted pattern. These xylem vessels are under considerable tension when drawing water up through the plant, and the strengthening of their inner walls prevents them from collapsing inwards.

Figure 2.29
Xylem and phloem

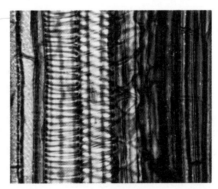

a Xylem LS (×240)

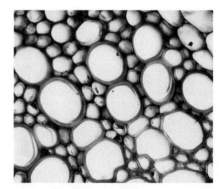

b Xylem TS (×260)

7 Draw a few xylem vessels in LS to show the various forms of thickening.

Phloem

All living parts of the plant must respire in order to produce the energy required for the life processes carried out by the cells. The products of photosynthesis, mainly sugars, must be transported from the leaves to the non-photosynthetic parts of the plant. In angiosperms, phloem tissue carries out this function. Like the xylem vessels, the phloem is made up of a vertical column of sieve-tube elements which are joined at their end walls; these end walls are called **sieve plates** and they have perforations which connect neighbouring cells. Lying beside the sieve-tube elements are **companion cells**

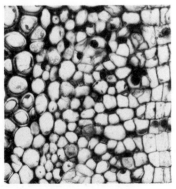

c Phloem TS (×160)

which are thought to provide nuclear functions and energy for the sieve tubes, since as the sieve-tube cell matures it loses its nucleus and many of its organelles. Phloem can be differentiated into **protophloem** and **metaphloem**.

In herbaceous dicotyledons, the xylem and phloem tissues are found closely associated in **vascular bundles** which form a ring in young dicotyledon stems. The ground tissue inside the ring of bundles is called the **pith** and the tissue outside the bundles the **cortex**. The xylem lies next to the pith, and the phloem next to the cortex. They are separated by the cambium from which new cells are produced.

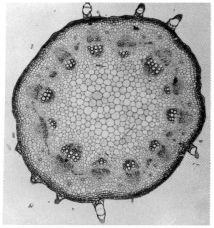 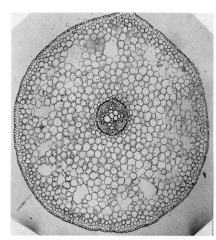

Figure 2.30   **a Dicotyledon stem TS (×17)**   **b Dicotyledon root TS (×49)**

8 Make a biological drawing of Figure 2.30(a), a TS of a herbaceous dicotyledon stem. Label and annotate it according to the details given above.

9 Make a biological drawing of Figure 2.30(b), a TS of a dicotyledon root. Label and annotate it. Make notes on how it differs from the stem you drew in the previous question.

2.2.3 Mammalian histology

As in plants, the division of labour between different types of animal cell has led to specialised cells lying close together as tissues, and carrying out the same function. These tissues are then structurally organised to form organs. This can be illustrated by looking at the situation in the mammal.

The tissues can conveniently be divided into four groups: **epithelial**, **connective**, **muscular** and **nervous**.

Epithelial tissue

Epithelial tissue is comparable to the epidermis of plants in that it protects underlying tissue from damage from abrasion or pressure, or invasion by disease-causing organisms. This is achieved by the epithelium either becoming thickened and toughened, or by a rapid replacement of cells by cell division so that any cells which are lost or damaged are quickly replaced. At the base of the epithelium is a network of collagen fibres called the **basement membrane**. Epithelia can be grouped according to the shape and arrangement of the cells.

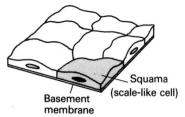

Figure 2.31 Squamous cells

Basement membrane

Squama (scale-like cell)

Classification based on the structure of the cells:

(i) **Squamous:** the cells are thin, flat and scale-like. The edges of the cells are irregular so that they fit together like crazy paving.

(ii) **Cuboidal:** the cells are shaped like a cube.

(iii) **Columnar:** the cells are tall and narrow.

(iv) **Ciliated:** the cells are usually tall and narrow (columnar) but with cilia on their free surfaces.

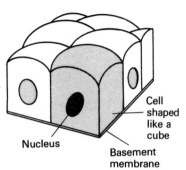

Nucleus

Cell shaped like a cube

Basement membrane

Figure 2.32 Cuboidal cells

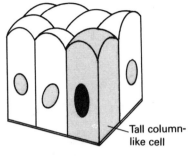

Tall column-like cell

Figure 2.33   Columnar cells

Cilia

Columnar cell

Nucleus

Basement membrane

Figure 2.34   Ciliated cells

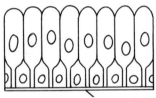

Basement membrane

Figure 2.35 Pseudostratified epithelium

Classification based on the number of cell layers:

(i) **Simple:** the epithelium is one cell thick.
Figures 2.31–2.34 are all simple epithelia.

(ii) **Pseudostratified:** the epithelium appears to be more than one cell thick although all the cells are in contact with the basement membrane.

(iii) **Stratified:** the epithelium consists of several layers of cells.

(iv) **Transitional:** the cells can change their shape, for example when they are stretched.

(v) **Glandular:** all glands are epithelial in origin and they are classified by the way in which they produce their secretions. **Exocrine glands** secrete to the free surface of the epithelium by ducts, **endocrine glands** have no ducts and secrete directly into blood vessels.

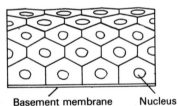

Basement membrane   Nucleus

Figure 2.36 Stratified epithelium

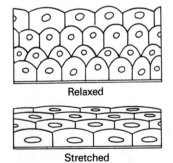

Relaxed

Stretched

Figure 2.37 Stratified transitional epithelium

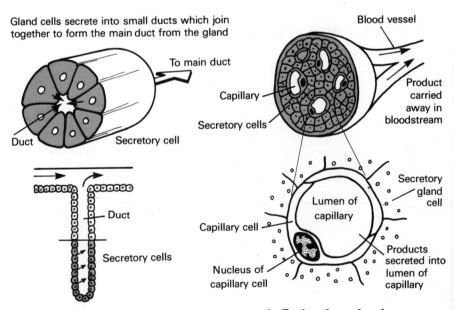

Gland cells secrete into small ducts which join together to form the main duct from the gland

To main duct

Duct   Secretory cell

Duct

Secretory cells

a Exocrine gland

Figure 2.38

Blood vessel

Capillary

Secretory cells

Product carried away in bloodstream

Capillary cell

Lumen of capillary

Nucleus of capillary cell

Secretory gland cell

Products secreted into lumen of capillary

b Endocrine gland

Connective tissue

Connective tissue binds together and supports all other types of tissue. It is a composite structure consisting of a variety of cells, several types of non-living fibre and an extracellular semi-fluid or fluid matrix. The cells are therefore usually widely separated from each other.

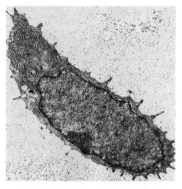

Figure 2.39
TEM chondrocyte (×6000)

Types of cell found in connective tissue

(i) **Fibroblasts:** these cells secrete the extracellular matrix which often includes various protein fibres.

(ii) **Fat cells:** these are cells which store fat, making up **adipose tissue**.

(iii) **Macrophages and mast cells:** these are defensive cells, macrophages enveloping infective organisms and mast cells bringing about inflammation by releasing histamine.

(iv) **Chondrocytes and osteocytes:** these are cartilage cells and bone cells respectively.

Fibres found in the extracellular matrix

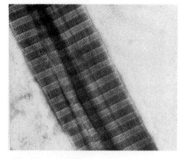

Figure 2.40
TEM collagen fibres (×82 000)

(i) **Collagen:** these are the most common fibres, giving strength and support to connective tissues. They consist of the protein **collagen** and are loosely packed in **areolar** tissues but closely packed in **white fibrous** tissue (found in tendons and ligaments).

(ii) **Reticular:** these fibres are made of the protein **reticulin**. The fibres are branched, and support the matrix and cells. They are commonly found in organs like the liver and the lymph nodes.

(iii) **Elastic:** these fibres are made of the protein **elastin**. It can also occur as thin sheets. These fibres have elastic properties, resuming their shape after having been stretched, and are found in blood vessels, lungs, and so on.

(iv) **Cartilage, bone, blood and lymph:** these are also connective tissues.

Arrangement of cells, fibres and extracellular matrix in different connective tissues

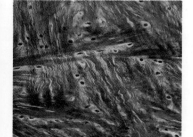

Figure 2.41
Areolar tissue (×120)

(i) **Areolar:** this is a loose connective tissue of widely separated fibroblasts lying in the extracellular matrix, which also contains randomly distributed collagen and elastin fibres.

1 Make a simple drawing of Figure 2.41 and identify and label all relevant structures.

(ii) **White fibrous**: this is a dense fibrous connective tissue made up of large numbers of bundles of collagen fibres packed closely together. The collagen fibres run parallel to each other and are interspersed with fibroblasts. This tissue is very strong and flexible but has tensile strength, resisting stretching. It is abundant in tendons, some ligaments, the sclera of the eye, the capsule of the kidney, cartilage and bone.

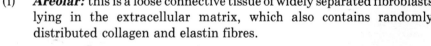

Figure 2.42
White fibrous tissue (×15)

2 Make a simple drawing of Figure 2.42 and label the relevant structures.

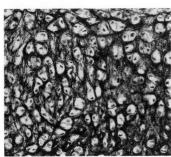

Figure 2.43
Yellow elastic cartilage
(×26)

(iii) **Yellow elastic:** in this tissue the yellow elastic fibres are branched and irregularly arranged to form a loose network. Some collagen fibres are also present and the fibroblasts are randomly scattered throughout the matrix. The elastic fibres are very flexible, returning to their original shape when stretched, while the collagen fibres give the tissue its strength. It is found in places such as the ligaments, artery walls and in the lungs.

3 Make a simple drawing of Figure 2.43 and label the relevant structures.

(iv) **Cartilage:** this is very hard and tough, but still flexible. It is the forerunner of bone, being laid down in the embryo and forming a template for the bony skeleton. The matrix is laid down by cells called **chondroblasts**. The matrix also contains many fine collagen fibres. The chondroblasts eventually come to lie in small spaces called **lacunae** at which time they are known as **chondrocytes**.

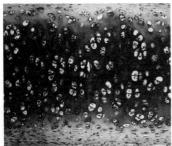

Figure 2.44
Hyaline cartilage (×64)

4 Make a labelled drawing of the chondrocyte in Figure 2.39, on the previous page.

Three main types of cartilage can be recognised, each of which has a special arrangement of components to suit its purpose.

Hyaline cartilage: this consists of a matrix of **chondrin** which is secreted by the small clusters of chondroblasts and it also contains collagen. It is very tough and is found covering the ends of the limb bones where they articulate at the joints; it is also found where the ribs are joined to the sternum. Rings of hyaline cartilage occur in the trachea and bronchi and prevent them from collapsing.

Yellow elastic cartilage: the matrix contains yellow elastic fibres arranged in a random fashion. It allows flexibility yet the structure still maintains its shape. It is found in the pinna of the ear and in the epiglottis.

White fibrous cartilage: this is very strong and capable of resisting stretching as the matrix contains many white collagen fibres. It is found in the intervertebral discs of the spinal column.

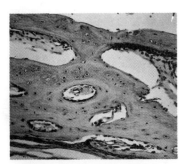

a Spongy bone (×64)

(v) **Bone:** this is the most common of skeletal materials. The bone cells, called **osteocytes** or **osteoblasts**, are embedded in a firm matrix which consists of about 30 per cent organic materials, mostly collagen fibres, and 70 per cent inorganic salts. Like cartilage, the bone cells are enclosed in **lacunae** which are connected to each other by a system of fine canals called **canaliculi**. Bone can withstand considerable compression and can resist tension.

Two main types occur:

Spongy bone: this occurs in the embryo and in the epiphyses (expanded ends) of long bones. It consists of a network of thin bony struts known as **trabeculae** which are arranged in the direction in which the bone is stressed.

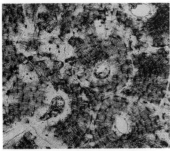

b Compact bone (×32)

Figure 2.45

Compact bone: the matrix consists of bone collagen, made by the osteoblasts, together with quantities of magnesium, sodium, carbonates, nitrates and hydroxyapatite. The resulting structure is immensely strong because the materials are laid down in a cylindrical series of layers (lamellae) suited to the forces acting on it. One cylindrical structure is known as a **Haversian system**, consisting of concentric lamellae. The matrix contains a system of

Haversian canals running parallel to the longitudinal axis of the bone. The canals contain nerves, blood and lymph vessels.

5 Draw and label both types of bone in Figure 2.45.

(vi) **Blood:** this consists of various specialised cells suspended in a fluid matrix called plasma. About 55 per cent of the total volume of the blood is plasma, 45 per cent being cells.

Plasma: this is a pale straw-coloured liquid consisting mainly of water. About 10 per cent of it is dissolved and suspended substances, for example serum albumin and globulins, enzymes, mineral ions, digestion and excretory products.

Erythrocytes: known commonly as red blood cells, these are small circular biconcave disc-shaped cells without nuclei. Their diameter is about 7–8 μm. Their function is to carry oxygen.

The other main cell in the blood is the *leucocyte* or white blood cell. There are various types:

Granulocytes: these white cells have a granular cytoplasm and a lobed nucleus and so are often called polymorphonuclear leucocytes. They can be subdivided into **neutrophils** (phagocytes), **eosinophils**, and **basophils** because of their different staining properties.

Agranulocytes: these have a non-granular cytoplasm. Two types are found – **monocytes**, which have a bean-shaped nucleus and are active phagocytes, and **lymphocytes**, which have very little cytoplasm and can be large or small.

Platelets: these are tiny membrane-bound cell fragments which help to initiate blood clotting. They have an irregular shape and there are about 250 000 per cm^3 of blood. They are also called thrombocytes.

6 Plate 2.3(a) is a TEM of a transverse section through a capillary. Draw, label and annotate.

Erythrocyte (section)

Neutrophil

– with drumstick chromosome

Eosinophil

Basophil

Small lymphocyte

Large lymphocyte

Monocyte

Platelets

Figure 2.46
Various blood cells

PLATE 2.3

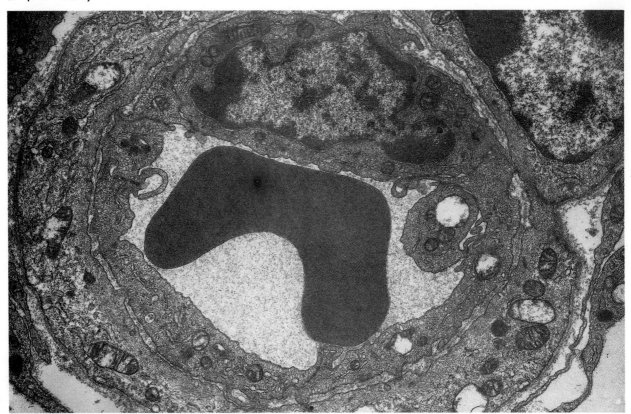

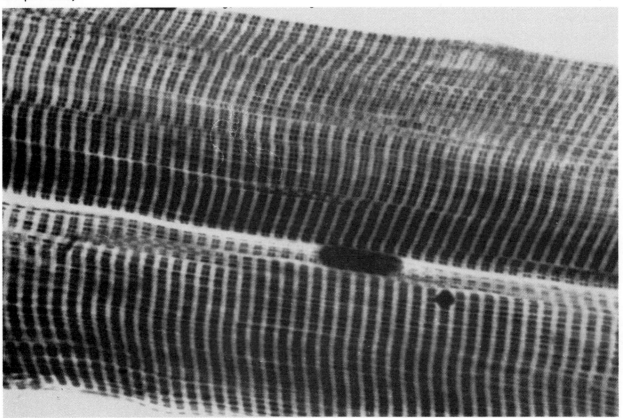

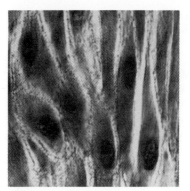

Figure 2.47
Smooth muscle (×73)

Muscular tissue

There are three types of muscle tissue, each structurally adapted to carry out a particular form of contraction.

(i) **Smooth muscle:** (also called visceral, unstriated, unstriped and involuntary). This consists of separate spindle-shaped muscle cells with single nuclei (uninucleate). The cells are held together by a connective tissue consisting mostly of collagen, to form a distinct muscle layer with the cells oriented parallel to each other. This type of muscle can maintain powerful contractions for long periods and has equally long relaxation periods; it is also capable of spontaneous rhythmic contractions. It is found in the walls of tubular organs and vessels and is involved in the movement of materials through them. The muscle cells are supplied with two sets of nerves, one from the parasympathetic system and one from the sympathetic system – providing accelerator and inhibitor facilities. It is found, for example, in the alimentary canal, the urinogenital tract, blood vessels, large lymph vessels and in the main ducts of glands.

7 Draw a few smooth muscle cells and label them.

(ii) **Skeletal muscle:** (also called striated, striped and voluntary). This consists of many small muscle fibres enclosed in bundles by collagen fibres and connective tissue. The bundles are attached to bone by tendons which consist almost entirely of collagen. Each muscle fibre is enclosed by a membrane known as the **sarcolemma**. The fibres are cylindrical and lie parallel to their neighbouring fibres. Each fibre is made up of a number of thin **myofibrils**, which have a pattern of dark and light bands which can be seen under the light microscope, but are

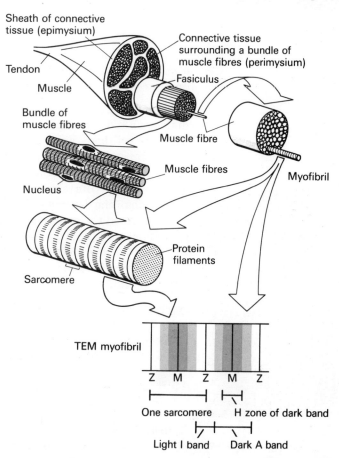

Figure 2.48
Skeletal muscle

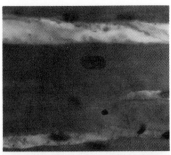

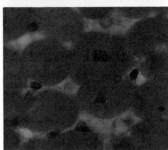

Figure 2.49
Cardiac muscle LS and TS, LM (×480)

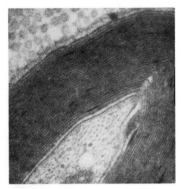

Figure 2.50
TEM myelin sheath (×36 000)

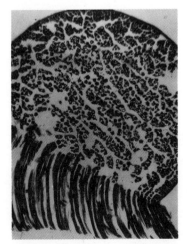

Figure 2.51
TS and LS nerve, LM (×58)

particularly noticeable in TEM micrographs. These are caused by the arrangement of **myosin** and **actin** filaments. Each myofibril consists of units called **sarcomeres**.

This type of muscle is capable of fast and powerful contraction but it readily becomes fatigued, requiring a resting period – the **refractory period** – before contracting again. It is the muscle that is used in moving the various parts of the body, particularly the limbs in locomotion. It is well supplied with blood vessels and contains mitochondria. It is innervated by the voluntary nervous sytem.

8 Make a labelled biological drawing of Plate 2.3(b), page 60.

(iii) **Cardiac muscle:** this type is only found in the heart. It is similar in appearance to skeletal muscle, having the same banding pattern, but the fibres are branched and connected with one another in such a way that impulses spread throughout the fibres in a co-ordinated manner. It has a long refractory period enabling it to recover completely from the previous contraction, so preventing fatigue. It is even more generously supplied with blood vessels and mitochondria than skeletal muscle.

9 Make labelled biological drawings of Plate 2.4(a) and (b).

Nervous tissue

This type of tissue is made up of densely packed and interconnected nerve cells called **neurones** whose function is to transmit impulses between receptors and effectors. It also contains **neuroglial cells** and connective tissue which is well supplied with blood vessels.

Neurones: these are cells which are capable of transmitting electrical impulses. There are three main types: **sensory** or **afferent** neurones which carry impulses from the receptors to the central nervous system (CNS – brain and spinal cord); **motor** or **efferent** neurones which carry impulses from the CNS to receptors; and **association** or **intermediate** neurones which link the sensory and motor neurones.

Each neurone consists of a **cell body** which can vary in size (3–100 μm in diameter) and contains the nucleus and most of the organelles. Extending from the cell body are a number of cytoplasmic processes. The processes which transmit impulses towards the cell body are called **dendrons,** while those which conduct them away from the cell body are called **axons**. The axons, or nerve fibres as they sometimes are called, are much longer and thinner than the dendrons. These nerve fibres may be surrounded by a fatty sheath termed the **myelin sheath** and are then known as **myelinated**, as in the cranial and spinal nerves; if they have no sheath, as in autonomic nerves, they are termed **non-myelinated**. The myelin sheath is formed by specialised cells called **Schwann cells** and is punctuated at intervals by **nodes of Ranvier**. Impulses are transmitted faster in myelinated nerves.

Nerves: these are made up of bundles of nerve fibres enclosed in a sheath of connective tissue termed the **epineurium**. Each nerve fibre is surrounded by connective tissue called the **endoneurium** and separating the individual nerve fibres is the **perineurium**.

10 From the description above, make a labelled biological drawing of Figure 2.51.

Neuroglia: this tissue is composed of **glial cells** which surround the neurones and support them, both in a physical and a physiological way.

PLATE 2.4

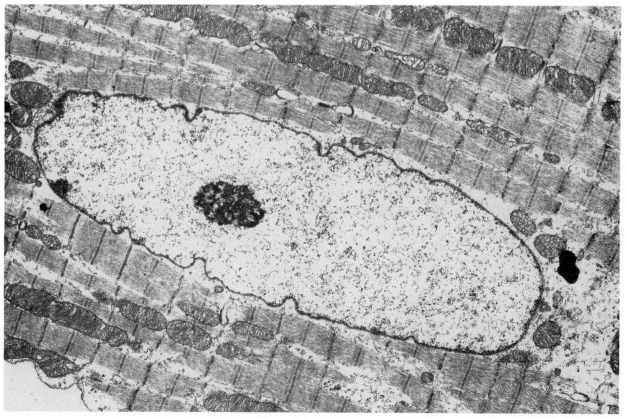

2.2.4 Lower animals

When drawing invertebrates we are usually looking at structures which have been magnified a few times at most.

The same general principles of drawing still apply. Proportions must be correct and to scale; distinctive structural features should be highlighted and the labelling must be correct. Explanatory annotations should finally be added.

Hydra

Plate 2.5(a) is a photograph of *Hydra*, a freshwater member of the phylum Coelenterata (Cnidaria). Plate 2.5(b) is a biological drawing and interpretation of *Hydra* from (a). It is drawn to the same scale as the photograph.

1 Examine the photograph and the biological drawing carefully and then make a suitable labelled and annotated biological drawing to the same scale of the *Hydra* shown in Plate 2.6(a). Note particularly any differences between Plate 2.6(a) and Plate 2.5(a).

2 Examine Plate 2.6(b). It shows a TS through *Hydra*. Make a labelled and annotated biological drawing of this light micrograph. Use reference books where necessary.

Drosophila

The fruit fly *Drosophila* has been used extensively in genetics experiments since the beginning of the twentieth century, and at present we probably know more about the genes of this little fly than of any other organism. *Drosophila* is ideal for genetic studies because:

(i)   it is small;

(ii) it has a short life cycle of about two weeks;

(iii) the female lays up to 200 eggs at a time;

(iv) various mutants are readily available;

(v)   it has only four chromosomes;

(vi) it can be fed on quite a simple diet.

3 Give reasons why these characteristics make *Drosophila* an ideal organism for genetic studies.

The normal or non-mutant fly is known as the **wild type**; it has long wings which extend beyond the abdomen, a light brown body colour and round red eyes.

With practice, the male can be distinguished from the female with the unaided eye, but it is easier at the beginning to observe them under a low-power stereomicroscope. When examining and counting the flies it is necessary to keep them stationary without causing them any harm. This can be achieved by temporarily anaesthetising the flies using ether in an 'etheriser'.

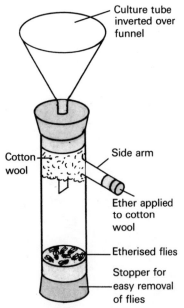

Figure 2.52
Drosophila *etheriser*

Culture tube inverted over funnel

Cotton wool

Side arm

Ether applied to cotton wool

Etherised flies

Stopper for easy removal of flies

PLATE 2.5

a *(×25)*

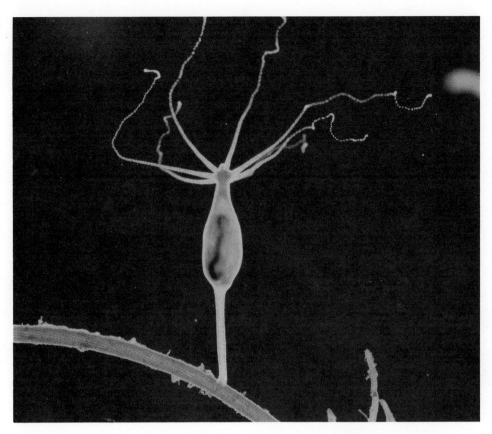

b

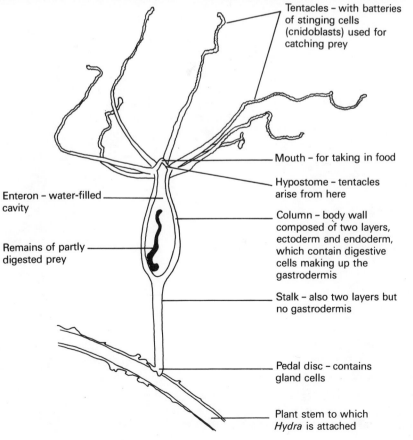

Tentacles – with batteries of stinging cells (cnidoblasts) used for catching prey

Mouth – for taking in food

Hypostome – tentacles arise from here

Column – body wall composed of two layers, ectoderm and endoderm, which contain digestive cells making up the gastrodermis

Enteron – water-filled cavity

Remains of partly digested prey

Stalk – also two layers but no gastrodermis

Pedal disc – contains gland cells

Plant stem to which *Hydra* is attached

PLATE 2.6

a *(×25)*

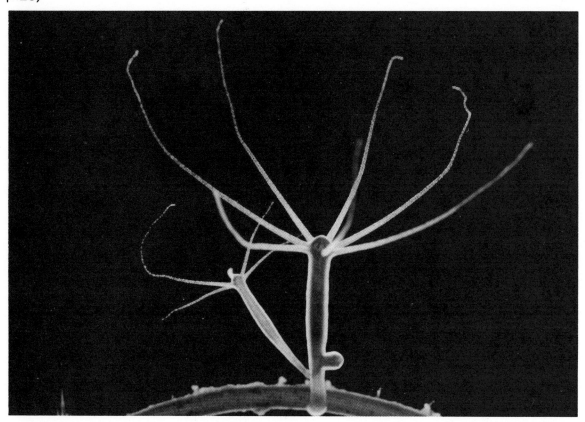

b *(×500)*

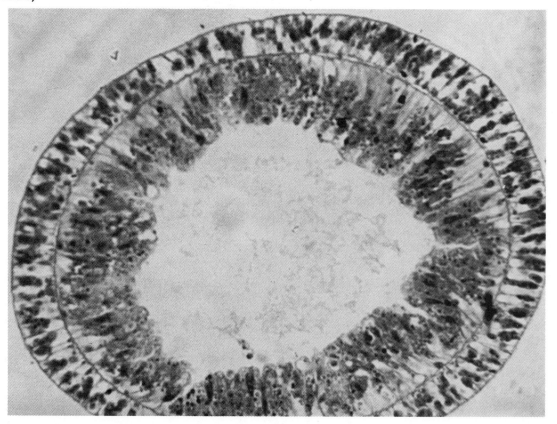

PLATE 2.7

a *(×8)*

b *(×100)*

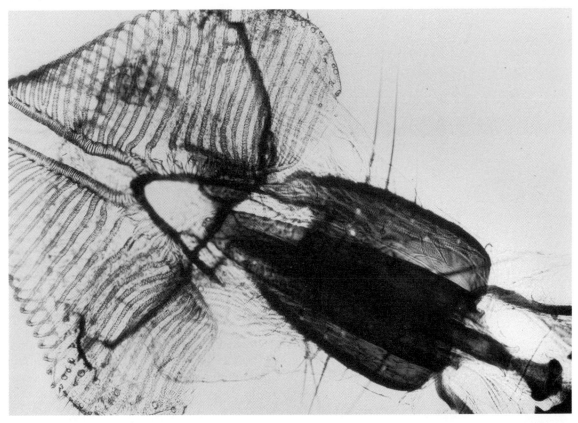

Actual size

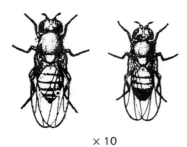

× 10

Figure 2.53
Drosophila:
actual size and ×10

The drawings in Figure 2.53 are of a male and a female wild type *Drosophila*, the dorsal view. Examine the drawings carefully and note the following differences between the sexes:

1. The female is larger than the male.

2. The tip of the female's abdomen is pointed whereas that of the male is rounded.

3. When viewing from the dorsal surface the tip of the male's abdomen is more heavily pigmented than that of the female.

4 Identify the male and female from the drawings and descriptions.

5 Make simple biological drawings of each sex of fly (×20). Label and annotate.

Mutants
Defective genes cause mutant forms of *Drosophila*, of which there are now a very large variety known. The most common and clearly recognisable are:

(i) Vestigial wing – the wings are small and imperfectly formed so that the fly is incapable of flight.

(ii) Ebony body – the body is more darkly pigmented than the wild type.

(iii) White eye – in contrast to the red eye of the wild type.

(iv) Bar eye – the eye is elongated instead of round.

(v) Curled wing – the wings are shorter than the wild type and curl up at the ends.

Figure 2.54
Various mutant Drosophila

A   B   C   D   E

6 Identify the mutations and sex of each of the flies (A to E).

Plate 2.7(a), on the previous page, is a close-up photograph of a hornet's head.

7 Make a labelled outline drawing of the hornet's head, to the same scale as the photograph.

Plate 2.7(b) is an LM of the proboscis of the house-fly *Musca domestica*.

8 Make a labelled biological drawing of the proboscis to the same scale as the photograph. Label and annotate it, paying particular attention to the relationship between structure and function; use whatever references are required.

2.3 APPLICATIONS OF DRAWING

Now that you have found out what is involved in recording your observations as labelled and annotated biological drawings, let us look at a number of different situations that you are likely to come across. Before you start any particular section, make sure that you have the knowledge and skills listed at the beginning of each exercise.

2.3.1 The cell – TEMs

Knowledge and skills required	Reference
Magnification and scales	1.1.9
How to draw a biological drawing	2.1.1
How to label a drawing	2.1.2
How to annotate a drawing	2.1.3
Structure of cell organelles	2.2.1
Functions of cell organelles	2.2.1

1 Examine Plates 2.8, 2.9, 2.10, 2.11, 2.12 and 2.13. They are TEMs of various types of cells. Some of the plates show only parts of cells while others contain a number of cells. Make biological drawings of each of the plates, using the same magnification for each.

2 Explain how you worked out the magnification in each case.

3 Label and annotate your drawings.

2.3.2 Protists

Knowledge and skills required	Reference
Magnification	1.1.9
How to draw	2.1.1
How to label	2.1.2
How to annotate	2.1.3
Structure of protists	Any suitable text

Protists are **unicellular** (single-celled) eukaryotes. Although they exist as single cells they can be quite complicated, with a variety of specialised structures.

The protists are classified as a kingdom which is divided into a number of phyla.

Plates 2.14, 2.15 and 2.16 contain photomicrographs of three protists:
  2.14 *Amoeba*
  2.15 *Paramecium*
  2.16 *Euglena*

Amoeba and *Paramecium* belong to the phylum Protozoa. Classification of this group is based on their method of locomotion –

(i)   by pseudopodia, for example Sarcodines;

(ii) by cilia, for example Ciliates;

(iii) by flagellae, for example Flagellates.

Euglena belongs to another phylum, the Euglenophyta, which is a small group of unicellular organisms which possess chloroplasts and move about by means of the flagellae.

1 Make biological drawings of the three protists.

2 Using other texts to enable you to identify the various structures, label your drawings.

3 Annotate the labels with particular reference to biological significance.

PLATE 2.8

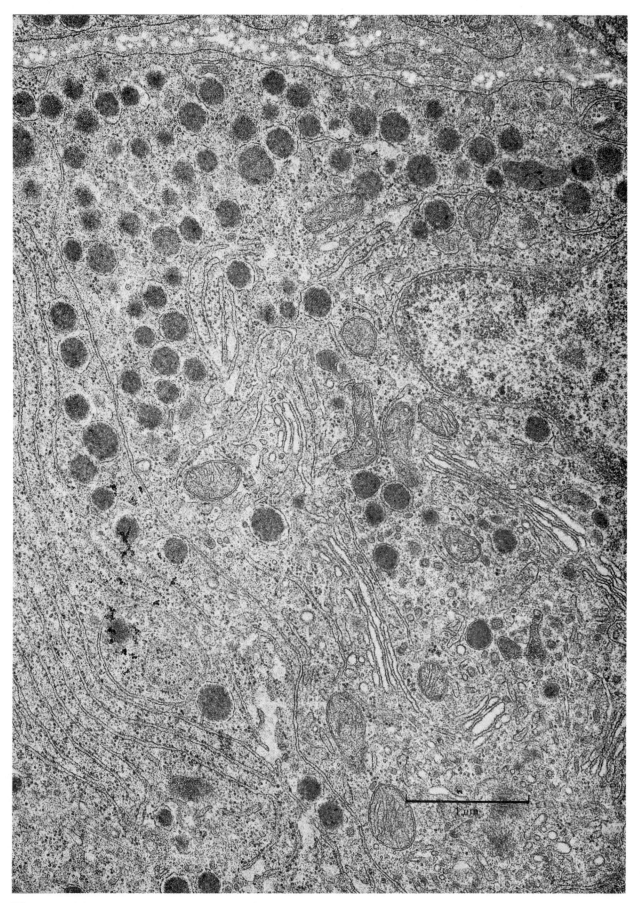

PLATE 2.9

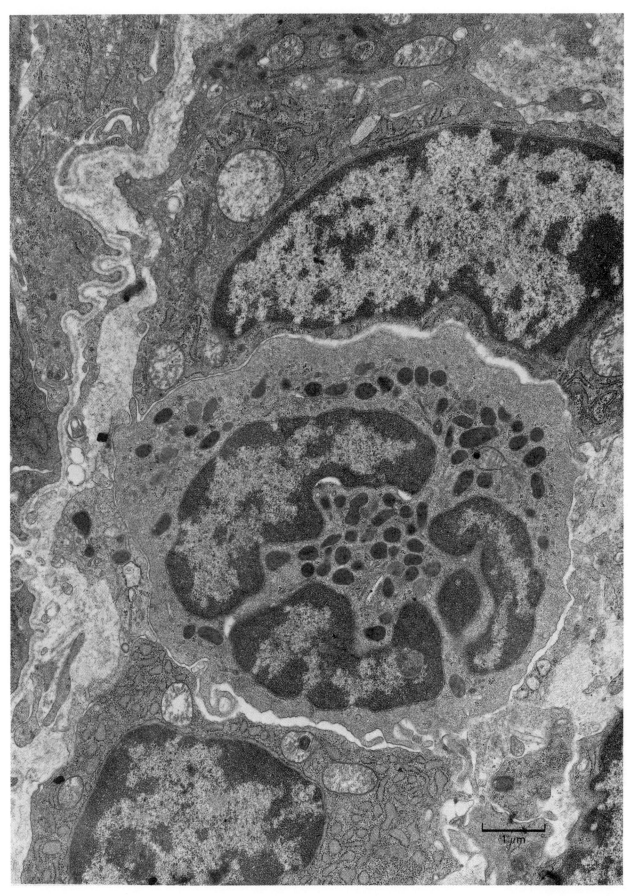

PLATE 2.10

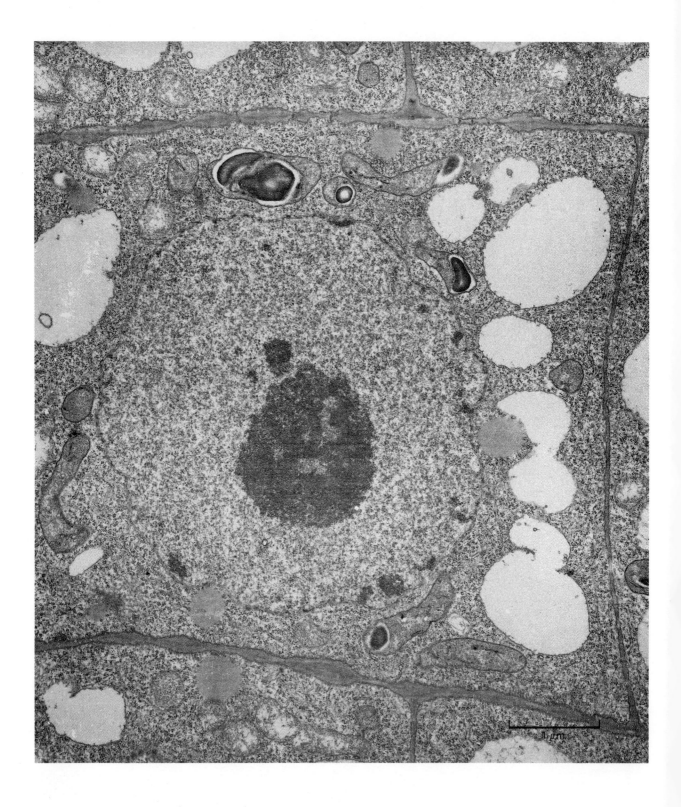

PLATE 2.11

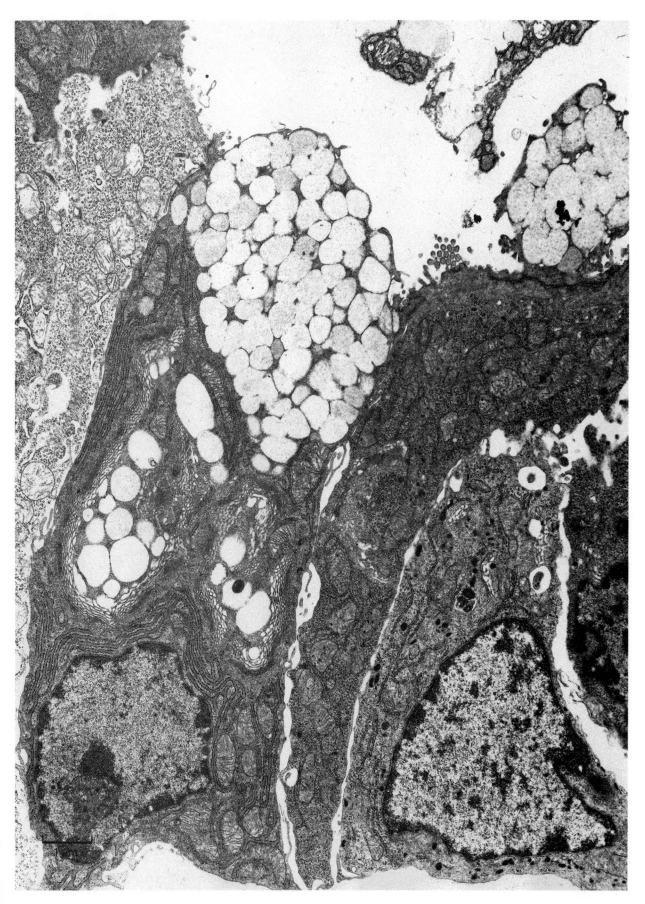

PLATE 2.12

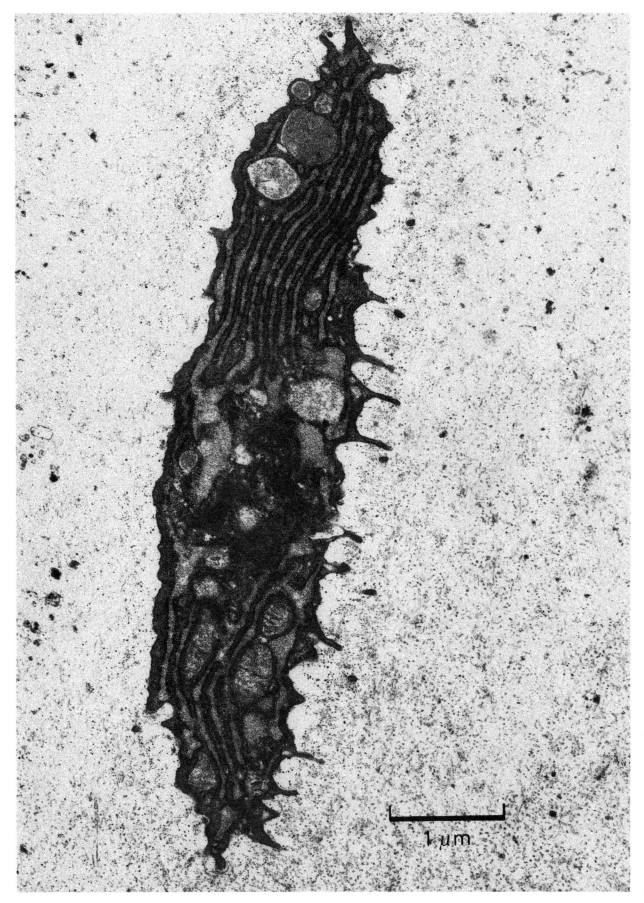

1 μm

PLATE 2.13

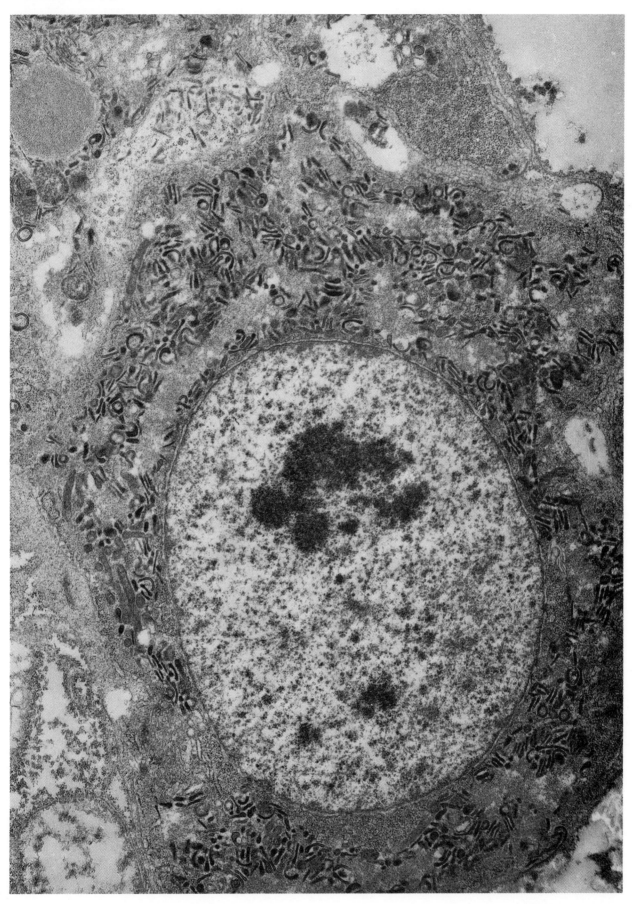

PLATE 2.14 *(×500)*

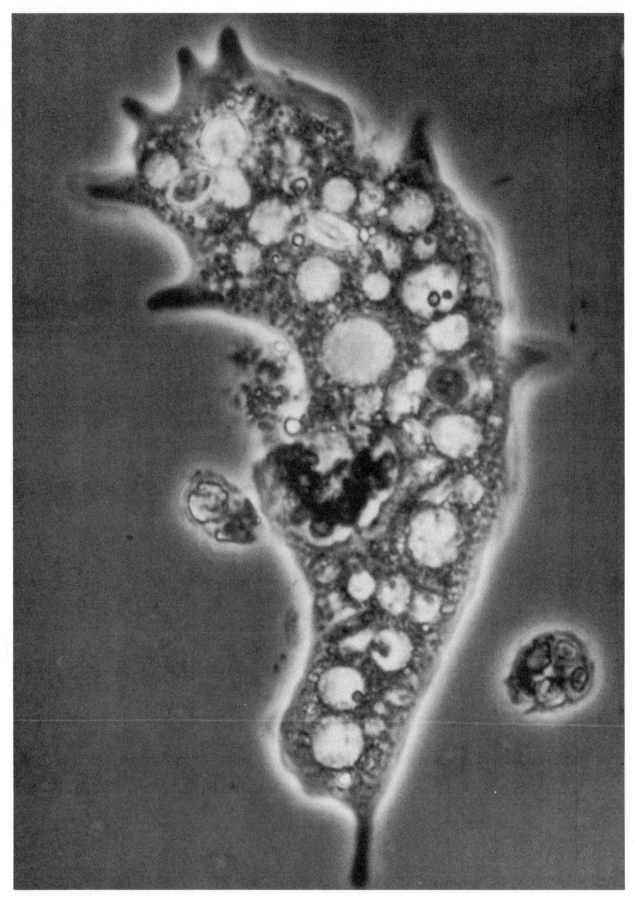

PLATE 2.15

(×1000)

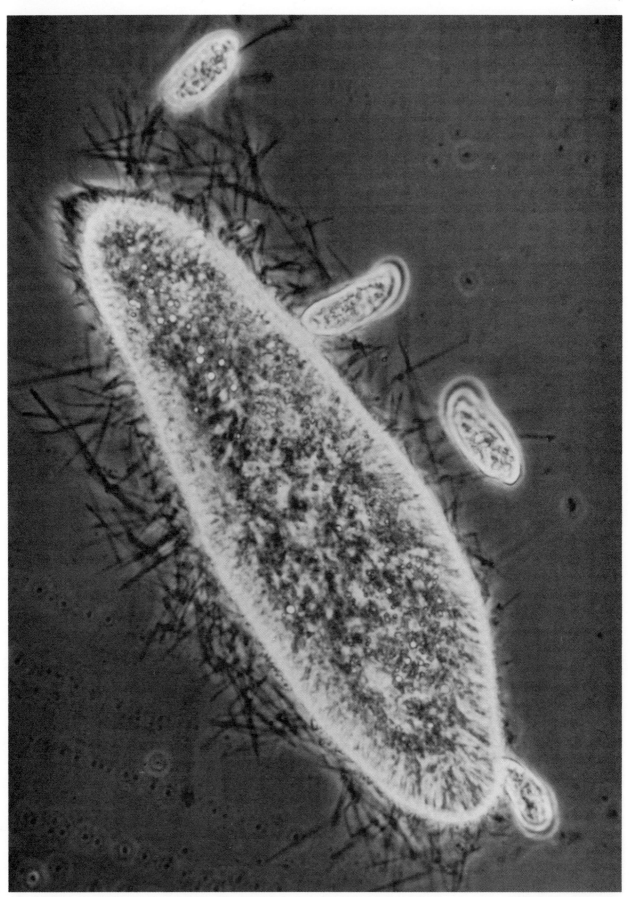

PLATE 2.16

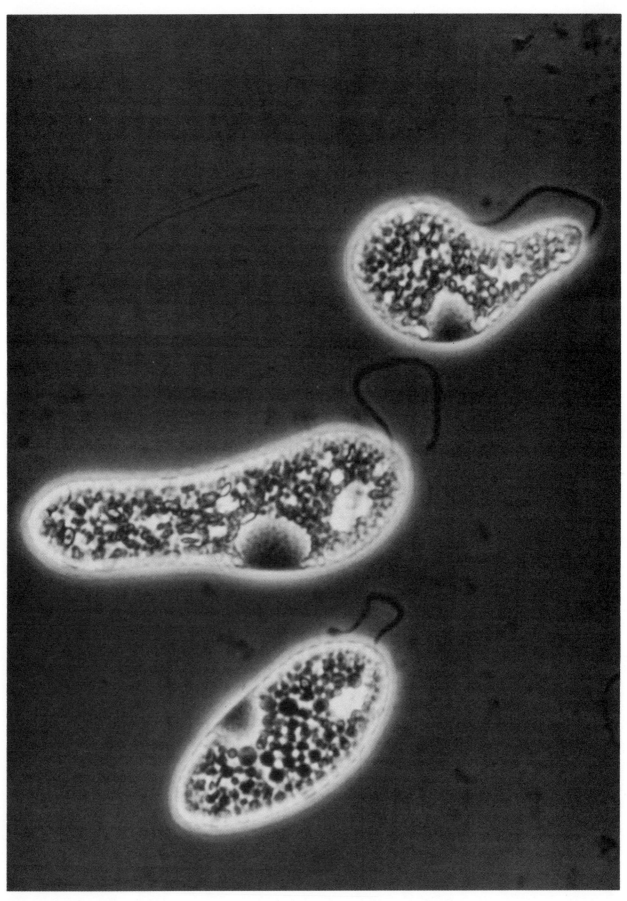

2.3.3 Angiosperm structures

Knowledge and skills required	Reference
Angiosperm histology	2.2.2; Suitable texts
Drawing	2.1.1
Labelling	2.1.2
Annotating	2.1.3

Primary structures

The parts of the plant which arise directly from the apical meristems and which increase the overall height of the shoot and the length of the roots are termed the *primary structures* of the plant. These make up the bulk of most plants and are separated structurally and functionally into three regions, the *roots*, *stems* and *leaves*.

The root

The root is one of the primary growth regions of a plant, and it is important to consider how the cells are formed and the pattern that they show when they have differentiated.

1 Examine Plate 2.17 which shows an LS through a bean root tip. Make a labelled biological drawing to show the relationships between the following: root cap, apical meristem, ground meristem, zone of elongation, zone of differentiation and any other tissues that you recognise. Use textbooks for reference.

In dicotyledons, the *radicle* which emerges from the seed continues to grow as the *primary* or *tap root*. Often, several lateral branches arise from the main root to produce the root system.

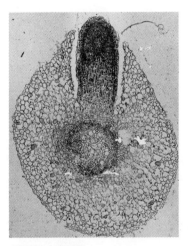

Figure 2.55
TS root with LS lateral root (×22)

Monocotyledons have a *fibrous* root system with no main root because the radicle dies soon after germination. Fine roots, known as *adventitious roots*, develop above the level of the first root.

The outermost layer of the root is the *epidermis*. Just above the growth zone this consists of cells which often possess finger-like extensions to their outer surface. To the naked eye these appear as root hairs; this region of the epidermis is called the *piliferous layer*.

2 Given that the function of these root hairs is to absorb water, how does their structure aid this process?

3 The root hairs are restricted to a relatively small region of the root epidermis. Suggest reasons for this.

As you have seen, the apex of the root is not protected by epidermal cells but by a thimble-shaped mass of parenchyma cells fitting over the meristem proper, called the *root cap*.

4 What function does this root cap have?

The *cortex* lies just inside the epidermis and consists of parenchyma cells. It is often the most important site for food storage in the plant. This is why we eat so many plant roots, for example carrots and parsnips.

PLATE 2.17

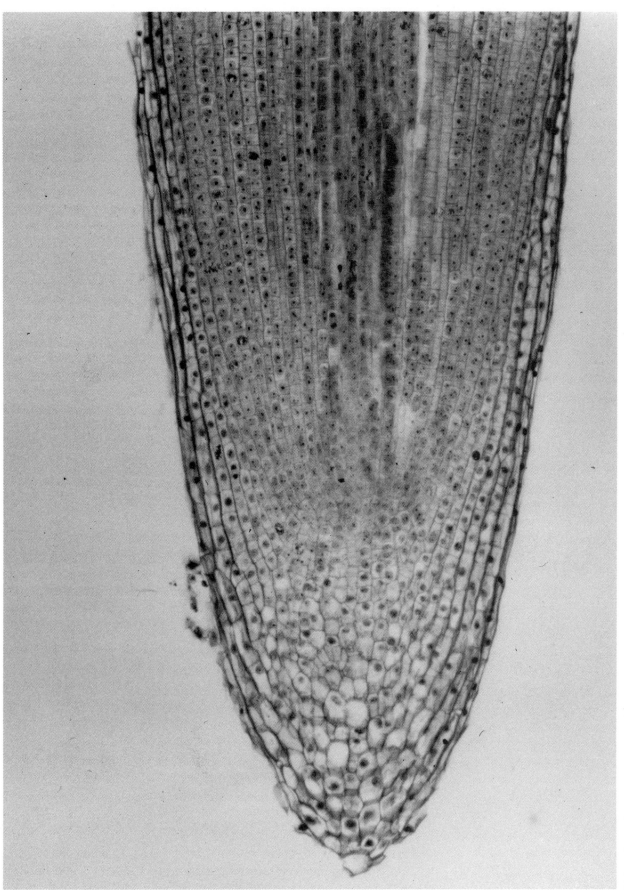

5 How would you test roots for a food storage substance and what would you expect to find?

The central core of vascular tissue, the **stele**, is bounded by a single layer of cells and is known as the **endodermis**. These cells have a band of waxy material called **suberin** in their radial and transverse walls. The effect of this band, known as the **Casparian strip**, is to give the plant total control over the passage of water into the vascular tissues of the root.

Within the endodermis are the vascular bundles, surrounded by relatively thin-walled undifferentiated **pericycle** cells which retain the ability to divide. The vascular tissues are made up of a central column of xylem, showing a star shape in TS, with the strands of phloem located between the rays of xylem.

6 Examine Plate 2.18 overleaf which shows a TS of a dicotyledon root, and make a detailed biological drawing.

7 Make a more detailed drawing of approximately a quarter of the stele of the root. Label and annotate it.

The stem

The stem is made up of basically the same structures and tissues as the root, but because they have to serve different functions, they are arranged in a different way.

Primary growth originates at the tip. The stem apex lies deeply enclosed and thus protected within the apical bud.

The outer epidermis is similar to that found elsewhere in the plant, being a single layer of tightly joined cells interrupted occasionally by stomata and multicellular hairs. Inside the epidermis is the cortex composed largely of parenchyma but frequently having an external layer, of collenchyma. The innermost parenchyma cells form a discrete layer, which often contains many starch grains and so is known as the **starch sheath**. The inner boundary of the cortex is marked by the vascular bundles which are arranged in a regular pattern around the circumference of the stem. Each vascular bundle has a complex structure, made up of at least four types of tissue. Externally a group of sclerenchyma fibres develop and within these are a series of phloem sieve tubes and companion cells. Inside the phloem are a few rows of thin-walled cambium cells, which appear roughly rectangular in TS and have a meristematic function. The innermost of the bundle tissues is the xylem.

The central portion of the stem is usually composed of large, thin-walled parenchyma cells and is termed the **pith**, but if the parenchyma cells stop growing while the rest of the stem continues to enlarge, a **pith cavity** is formed.

8 Examine Plate 2.19 (page 83) which is a TS of a stem of the dicotyledon *Helianthus* stem. Draw a sector of the stem to show the positions of the different tissues and label it. Also show details of a few cells from each tissue.

9 Examine Plate 2.20 which is a TS of a stem of the monocotyledon *Zea* (maize). Draw a small sector to show the distribution of the tissues.

10 Make out a suitable table summarising the differences between the two types of stem.

PLATE 2.18

(×500)

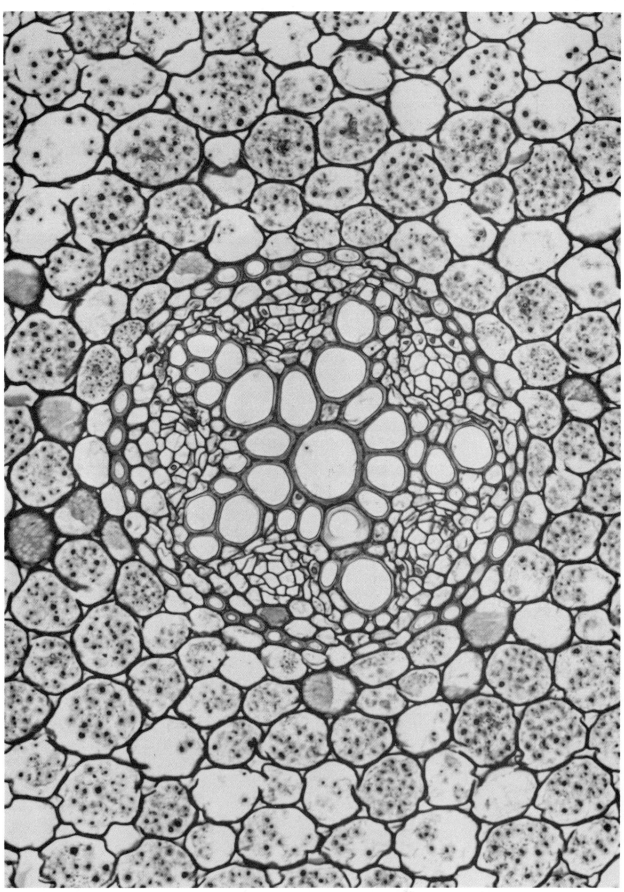

PLATE 2.19 (×55)

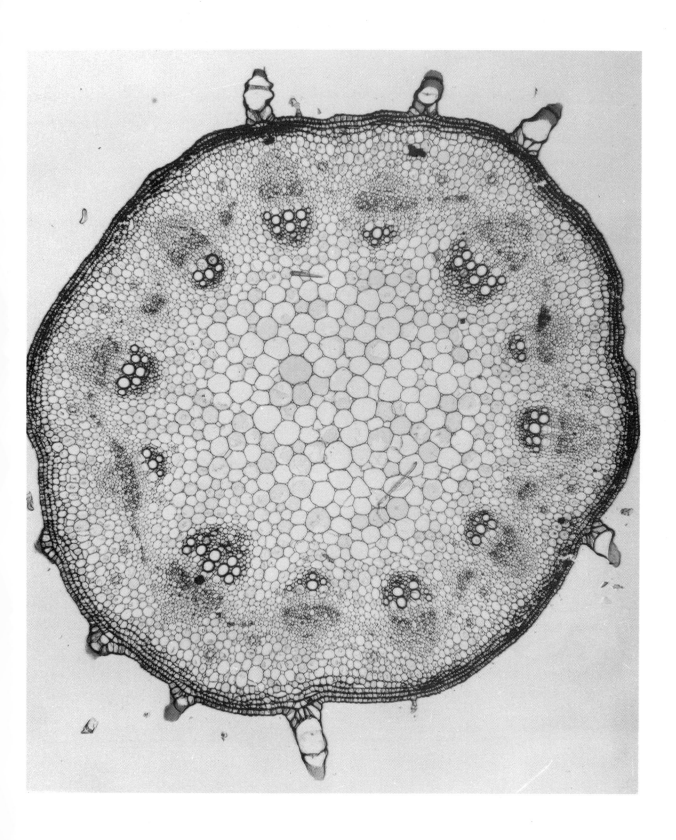

PLATE 2.20

(×23)

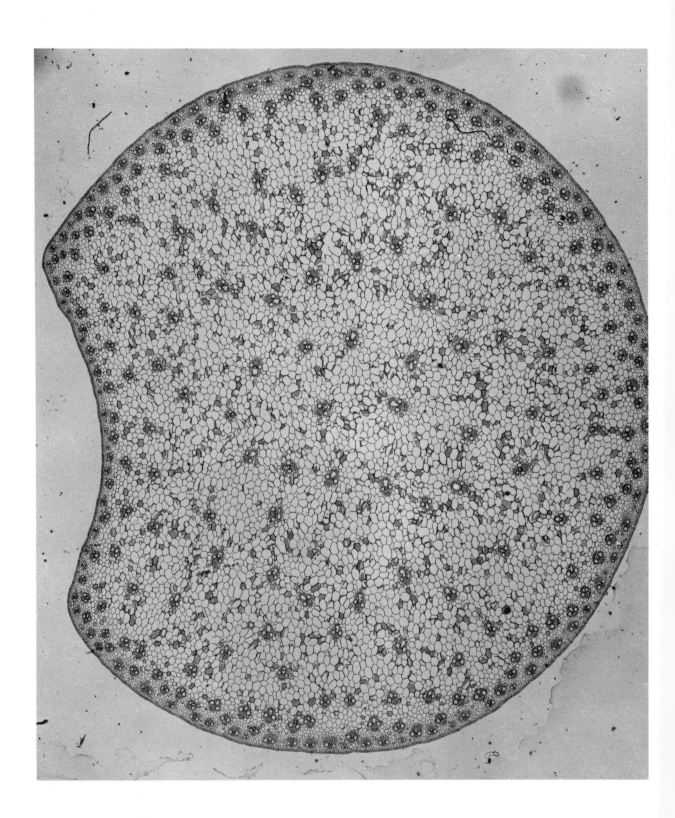

The leaf

The leaf consists of a lamina or blade, 10–15 cells thick, connected to the stem by the **petiole** or leaf stalk. The dicotyledon leaf is supported structurally and functionally by a network of branching veins arising from the **mid-rib**. Each vein is a vascular bundle.

The lamina is bounded by an upper and lower epidermis which have several functions:

1. They provide support by binding all the cells together, and hence counteract the hydrostatic pressure of the turgid mesophyll cells.

2. They allow light to penetrate into the leaf.

3. They prevent excess water loss from the leaf by developing a **cuticle**, at least on the upper surface.

4. They allow gaseous exchange and control water loss from transpiration through the stomata.

Internally, the leaf **mesophyll** consists of chlorenchyma cells (parenchyma with chloroplasts) arranged in two distinct layers. Immediately below the upper epidermis is a layer of **palisade** cells, oriented with their long axes at right angles to the surface of the leaf. Positioned in this way and containing high densities of chloroplasts, palisade cells ensure the maximum incidence of light on to the photosynthetic membranes of the chloroplasts.

Between the palisade layer and the lower epidermis lies the **spongy mesophyll**, so-called because about 50 per cent of this layer consists of **air spaces** lying between the cells. The arrangement of the cells provides a large surface area, which remains moist by transpiration and is typical of many respiratory surfaces.

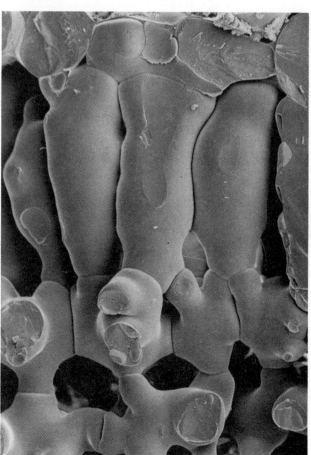

Figure 2.56
Freeze-fractured SEM of red kidney bean leaf (×830)

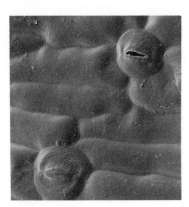

Figure 2.57
SEM open and closed
stomata of a dicotyledon
(×2600)

The lower epidermis in particular possesses a large number of **stomata**, which allow for gaseous exchange between the air spaces in the leaf and the atmosphere. Each **stoma** consists of two curved **guard cells** surrounding the **stomatal pore**. The guard cells are the only cells in the epidermis which contain chloroplasts.

Plate 2.21 is a TS of a dicotyledon leaf.

11 From the information given, make a labelled, annotated biological drawing of a representative portion of Plate 2.21.

Secondary structures

The secondary structures are involved in **secondary growth**, the process by which plants increase the girth of the stem and the roots. It occurs mainly in woody plants such as shrubs and trees but also occurs in herbaceous plants. These secondary tissues do not derive from the apical meristems as do the primary structures, but are produced by the **lateral meristems**.

The first stage in secondary thickening involves the cambium within the vascular bundles, the **fascicular cambium**, becoming joined together in between the vascular bundles by the formation of a new cambial layer, the **interfascicular cambium**. In this way a cylinder of meristematic tissue is formed. This now produces new xylem (**secondary xylem**) on the outside of the primary xylem and new phloem cells (**secondary phloem**) on the inside of the primary phloem.

As the girth of the stem increases, the cortex on the outside gets stretched. It is prevented from splitting by the development of a **cork cambium** which produces a layer of **cork** on the outside, thus keeping the stem sealed.

As the xylem tissue ages, it becomes increasingly lignified and the cells lose their contents. Rows of parenchyma cells known as **medullary rays** link the living cells in the cortex with the cells of the pith so that the transport of essential materials is not hindered.

In woody stems, practically all of the vascular tissue is secondary tissue. Each year additional secondary xylem and secondary phloem are added so that the characteristic tree rings are formed.

12 Examine Plate 2.22 (page 88) which shows a TS through a woody lime (*Tilia*) stem. What is the maximum age of this tree?

13 Make a biological drawing of Plate 2.22. Label and annotate it, using the information given above, as well as from other sources.

2.3.4 Hydrophytes and xerophytes

Knowledge and skills required	Reference
Angiosperm histology	2.2.2
Angiosperm structures	2.3.3
Drawing	2.1.1
Labelling	2.1.2
Annotating	2.1.3

In many habitats groups of plants are found which have many structural and physiological similarities, even though they are not closely related. The habitat selects for a few characteristic traits in all the populations, causing a form of convergent evolution.

PLATE 2.21 *(×240)*

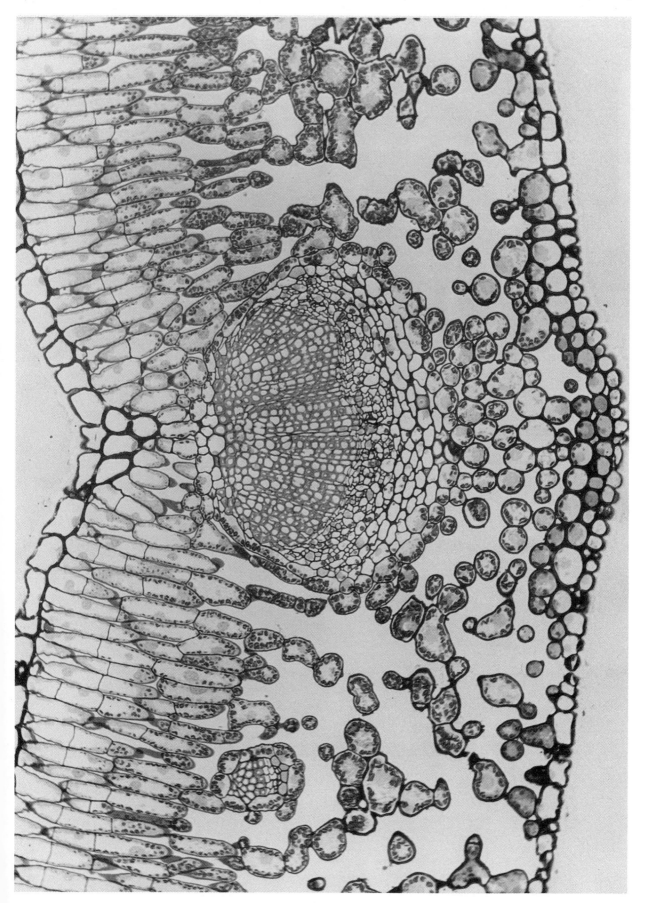

PLATE 2.22 *(×27)*

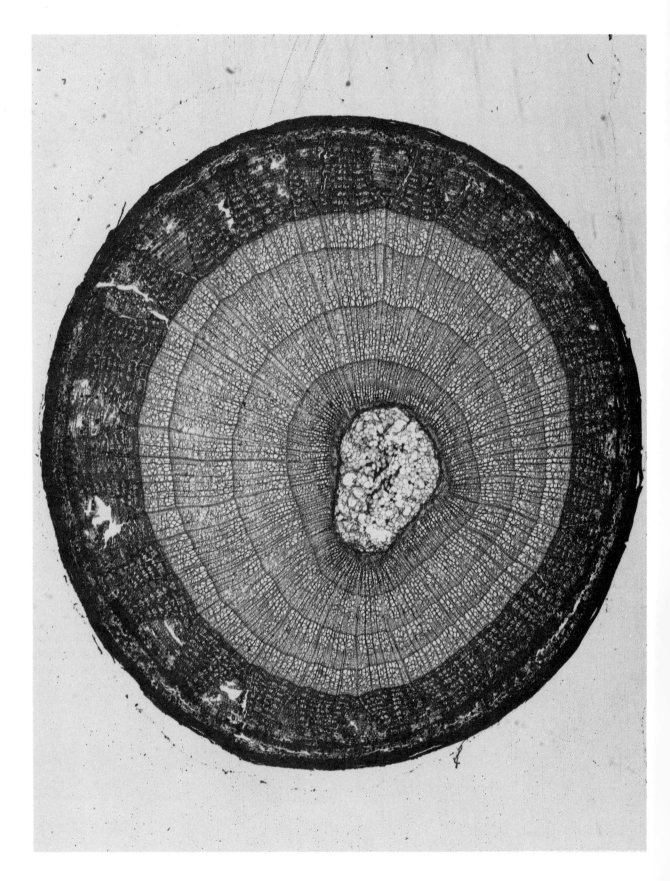

Cacti are not found in Africa, but its deserts include stem succulents from other families which resemble cacti in many features. The mangroves which inhabit tropical mudflats are taxonomically diverse, but share the same mode of life, having evolved very similar structural features.

An environmental factor which has one of the greatest effects on plants is the amount of water available to them. Plants can be divided into three categories depending on the amount of available water. The vast majority of plants are **mesophytes**, which for most of their lives have ample water. At the two extremes, **xerophytes** live in habitats where there is minimal water, so that it must be conserved at all times; **hydrophytes** live in conditions of excess water, existing on top of, or submerged in, water.

Xerophytes

A wide variety of adaptations, structural and otherwise, have evolved enabling plants to survive in conditions of minimal water. These include:

1. Adjustment of the life cycle so that during hot, dry periods the plant exists as a spore or a seed. Germination and maturation occur quickly when water becomes available.

2. The growth of an extensive or deep root system. Deep roots penetrate to the layers of the ground where water is more plentiful. Plants which have an extensive but shallow root system are able to absorb much of the small amount of water which falls on the surface of the ground before it evaporates.

3. Storing the water that they do absorb in large parenchyma cells, usually in the stem (in succulents).

4. Reducing the amount of water which is lost from the plant by structural modifications, such as:

(i)   reducing the leaves to protective spikes;

(ii)   reducing the number of stomata;

(iii) positioning the stomata in pits below the general leaf surface. Air in the pits becomes humid, thus reducing transpiration;

(iv)   covering the leaf with hairs, particularly around the stomata. This traps a layer of humid air and reduces transpiration;

(v)   covering the leaf epidermis with a thick waxy cuticle;

(vi)   folding or rolling the leaf so that a volume of humid air is trapped;

(vii) reversing the normal rhythm of stomatal opening so that they close during the day and open at night, when less evaporation of water is likely to take place.

Water problems are not only associated with high temperatures. Plants such as *Pinus* (Scots pine) and *Erica* (heather) have many xerophytic adaptations, yet they do not experience very high temperatures and there is usually a fairly high rainfall in their habits.

1 Why do plants like heather and the Scots pine suffer from a shortage of water? How do temperate broad-leaved trees overcome this problem?

2 Examine Plate 2.23(a), overleaf, which is a photomicrograph of a TS of *Ammophila arenaria* (marram grass), the coarse grass which colonises sand dunes. Make an annotated drawing summarising the adaptive features of the plant.

Hydrophytes

Plants which are submerged in water have difficulties both with gaseous exchange and with water currents. However, water is a dense medium and so it provides support for those parts of the plant that are submerged or floating.

PLATE 2.23

a *(×75)*

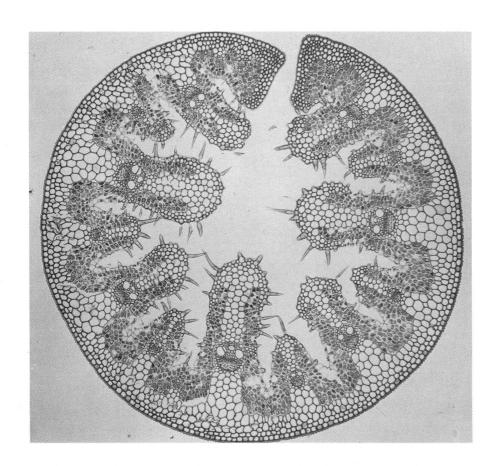

b

PLATE 2.24

a *(×75)*

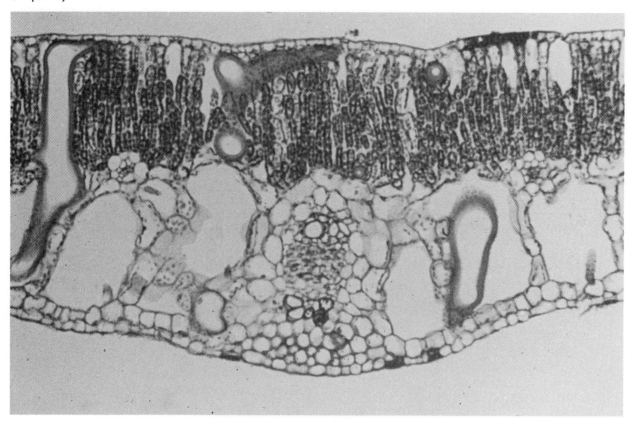

b *(×17)*

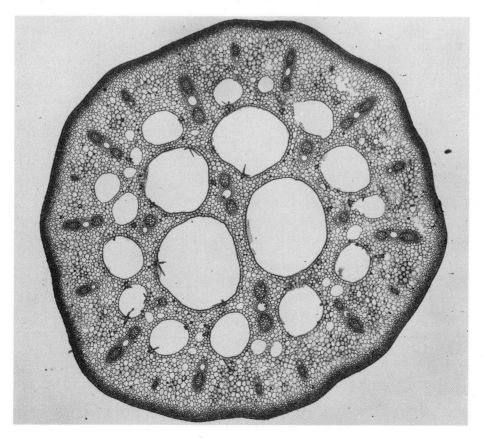

3 Examine the photograph of a typical hydrophyte on Plate 2.23(b), page 90. Make an annotated drawing summarising the adaptive features of the plant.

4 Plate 2.24 shows sections through the leaf and stem of the hydrophyte. Make suitable drawings of each, noting in particular any hydrophytic adaptations.

2.3.5 Primroses

Knowledge and skills required	Reference
Flower structure and pollination	Suitable text
Drawing	2.1.1
Labelling	2.1.2
Annotating	2.1.3

Primroses exist in two distinct forms in roughly equal numbers. These forms are known as *pin* and *thrum* and the two flowers are shown in Figure 2.58.

Figure 2.58
Pin and thrum primroses

Plate 2.25 shows the two morphological forms in LS.

The pollen grains of the primrose are a round shape. One of them is shown highly magnified in Figure 2.59.

The surface of the stigma, and the pollen grain of each type are shown together in Figure 2.60 (pin) and 2.61 (thrum). They are highly magnified, and at the same scale.

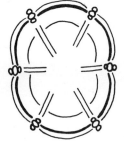

Figure 2.59
Pollen grains of the primrose Primula vulgaris

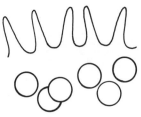

Figure 2.60
Pin stigma surface and pollen

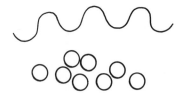

Figure 2.61
Thrum stigma surface and pollen

1 Make biological drawings of the two flower sections shown in Plate 2.25.

2 Label the drawings.

3 Annotate the drawings to indicate how the flowers are adapted for insect pollination.

4 Indicate how you think pollination works in the primrose and why there are two distinct forms.

PLATE 2.25

a *(×15)*

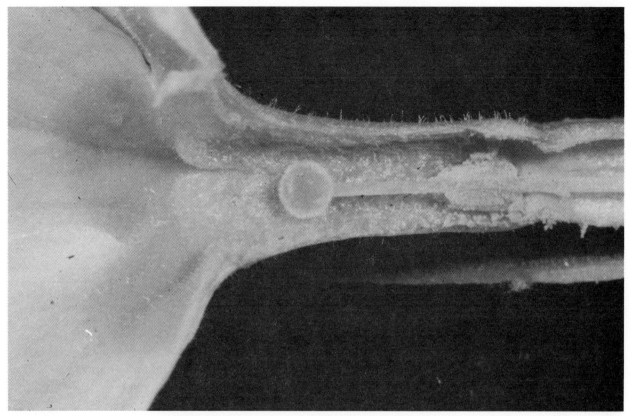

b *(×15)*

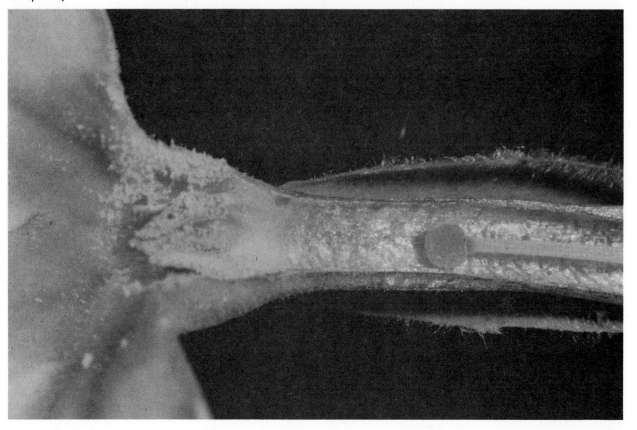

2.3.6 Parasites

Knowledge and skills required	Reference
Trematode and cestode structure and life cycles	Suitable text
Parasites	Suitable text
Drawing	2.1.1
Labelling	2.1.2
Annotating	2.1.3

There is a whole range of different associations which are found between organisms, including commensalism, phoresis, symbiosis, mutualism and parasitism.

A *parasite* can be defined as an organism that lives in or on another organism of a different species and derives nutrients from it without giving it any benefit.

Parasites are unique organisms in many ways since they have had to overcome certain difficult problems. The main adaptations to their problems are as follows:

1. *Degeneration:* parasites often show total loss of certain organs and mechanisms when they are compared with closely related non-parasitic species. Endoparasites tend to show greater loss, often having a lack of sense organs, reduced motility and a minimal nervous system. Gut parasites, like the tapeworm, have no digestive tract.

2. *Protective devices:* gut parasites have to avoid being digested by their hosts' enzymes and so have evolved a thick outer cuticle, which they coat in substances which locally inactivate the enzymes and also protects against the immune system of the host.

3. *Penetrative devices:* most endoparasites have the primary problem of getting inside the host. In animals this usually means that the parasite produces specific enzymes which digest the wall of the host, while in plants the parasites often produce cellulase enzymes.

4. *Attachment organs:* these are common in most parasites; in tapeworms they take the form of hooks and suckers.

5. *Dispersal mechanisms:* the most important problem for parasites is how to reproduce successfully and disperse. Many parasites which are unable to infect the primary host species directly, do so via one and sometimes two, other species – the *secondary* or *intermediate* hosts. The probability of infecting a host is increased in parasites by their prodigious powers of sexual and asexual reproduction.

1 Examine carefully the micrographs in Plates 2.26, 2.27 and 2.28.

2 Make a biological drawing of each of the specimens and with the aid of other reference material, label and annotate each drawing, in particular relating structure to function.

PLATE 2.26

a *(×25)*

b *(×18)*

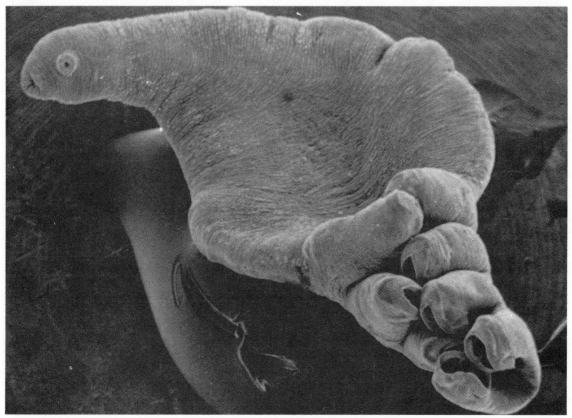

PLATE 2.27

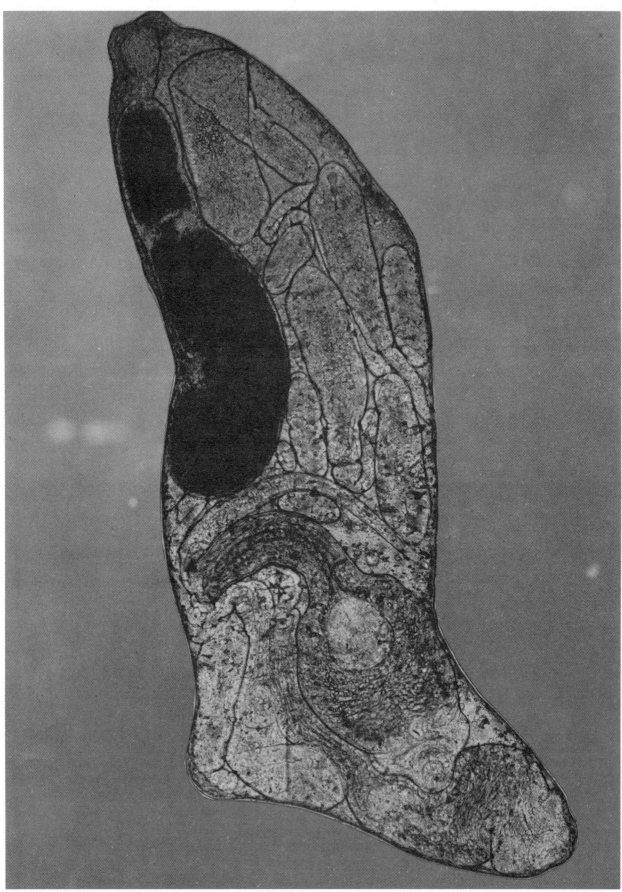

PLATE 2.28

a *(×73)*

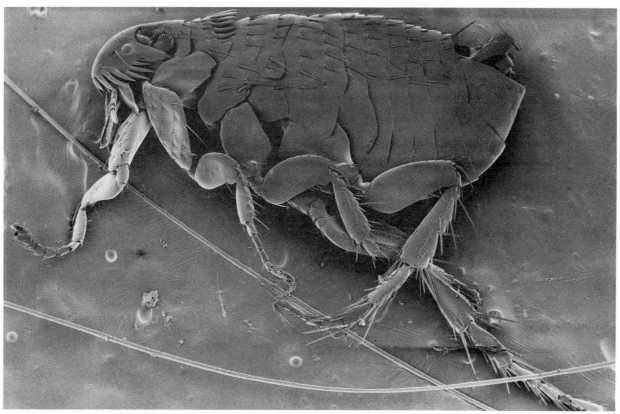

b *(×75)*

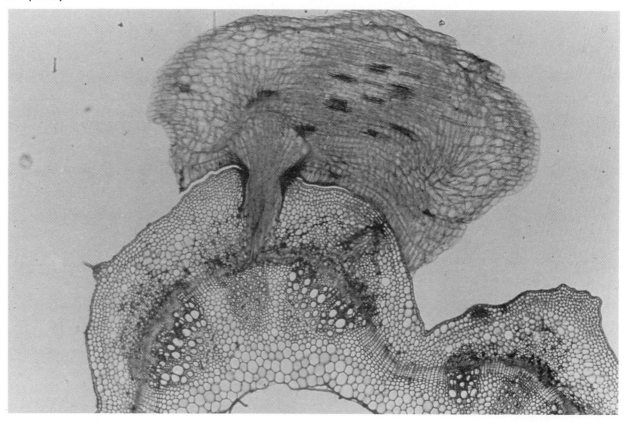

2.3.7 Mammal organs

Knowledge and skills required	Reference
Mammalian histology	2.2.3 and suitable texts
Microscopy	1.1.6
Drawing	2.1.1
Labelling	2.1.2
Annotating	2.1.3

You have already found out in Section 2.2.3 that cells of the same type are grouped together to form tissues which in turn are arranged in particular ways to form organs. Organs are further arranged into systems. Let us look at representative organs from a few human systems. We will mainly be concerned with analysing LM, TEM and SEM micrographs and representing them in the form of biological drawings, completed by labels and annotations.

Alimentary canal (digestive system)

The alimentary canal can be considered to be a tube of variable width, running from the mouth to the anus. Although there are differences between the regions of the gut, the wall is essentially made up of four layers: the **mucosa**, the **submucosa**, the **muscularis externa** and the **serosa**.

1. **Mucosa:** this is the innermost layer which in turn consists of three layers:

(i)   The epithelium – this lines the lumen of the gut and consists of a layer only one cell thick. The type of cell making up the epithelium varies from region to region.

(ii) The lamina propria – a layer of loose connective tissue supporting the epithelium; it can also contain glands, blood vessels and lymph vessels.

(iii) The muscularis mucosa – a thin layer of smooth muscle fibres.

2. **Submucosa:** this consists mainly of collagen fibres and elastic fibres; it also contains blood vessels, lymph vessels, nerves and glands.

3. **Muscularis externa:** this external muscle coat consists of an inner layer of fibres running around the canal in a circular arrangement, and an outer layer of fibres running lengthwise along the canal.

4. **Serosa:** this outer coating consists of areolar tissue which is continuous with the mesenteries supporting the gut.

Plates 2.29, 2.30 and 2.31 contain micrographs of various parts of the alimentary canal.

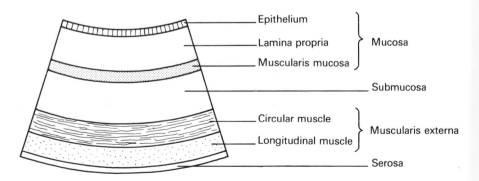

Figure 2.62
Layers in the gut wall

1 Using the information given previously in this section, and other reference books, identify the micrographs, make biological drawings of each, and label and annotate them. Estimate the magnifications in each case and state what type of microscope was used to obtain them.

2 Make out a table listing the characteristic features of each section of the alimentary canal.

3 Describe how the structure of each region is related to the functions that it carries out.

Plate 2.32 contains micrographs of various organs which are associated with the digestive system.

4 Using available sources, identify these micrographs, draw, label and annotate them.

PLATE 2.29

a *(×40)*

b *(×40)*

c *(×400)*

d *(×40)*

e *(×7200)*

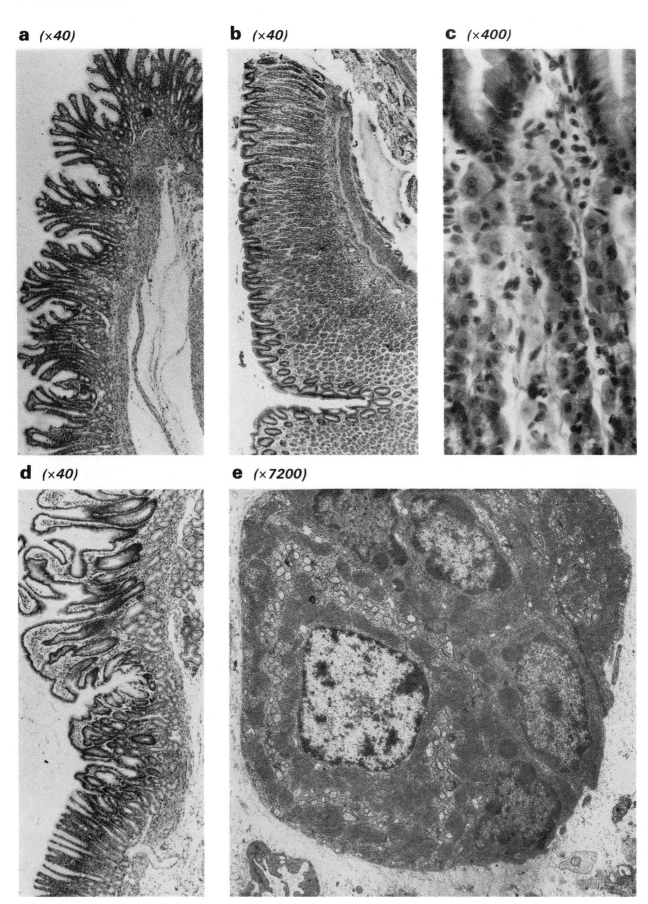

PLATE 2.30

a *(×600)*

b *(×500)*

c *(×60 000)*

d *(×40)*

e *(×40)*

f *(×250)*

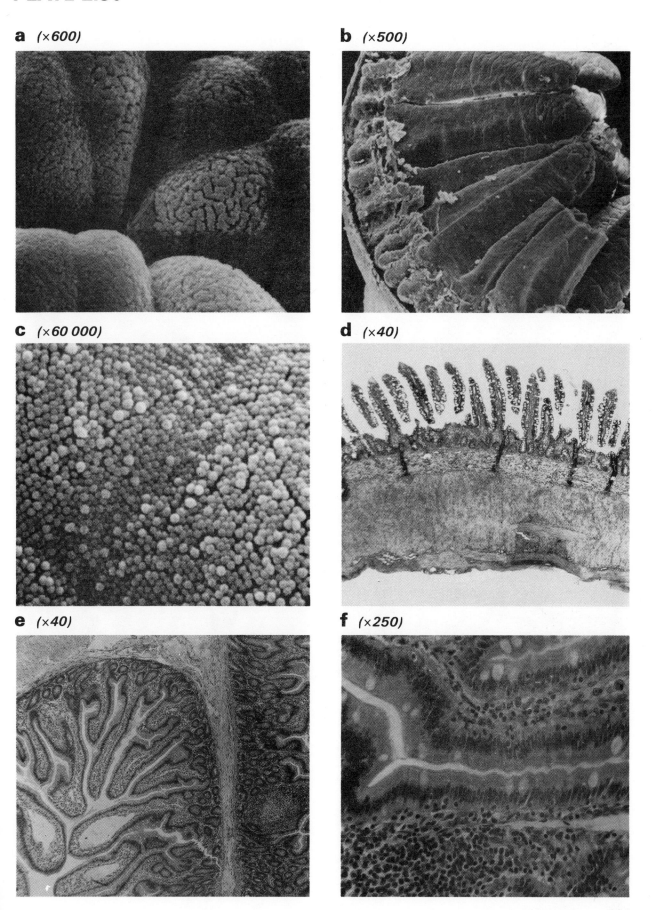

PLATE 2.31

a *(×130)*

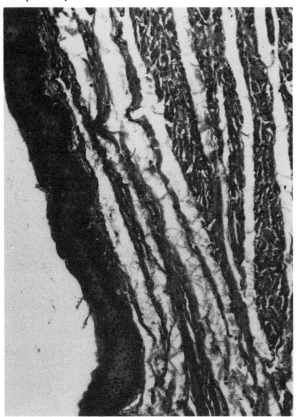

b *(×40)*

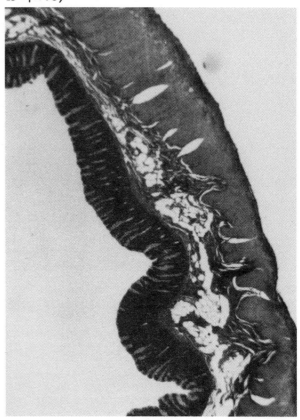

c *(×100)*

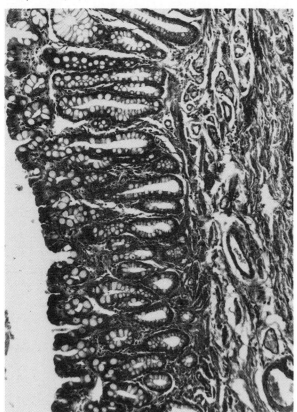

d *(×50)*

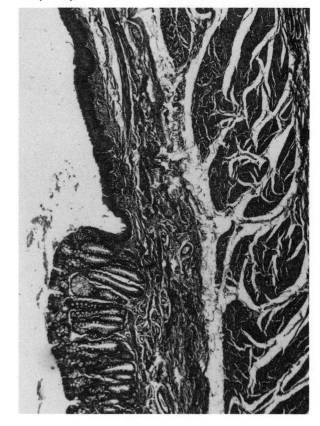

PLATE 2.32

a *(×100)*

b *(×40)*

c *(×40)*

d *(×40)*

e *(×1000)*

f *(×250)*

g *(×1000)*

h *(×270)*

i *(×4250)*

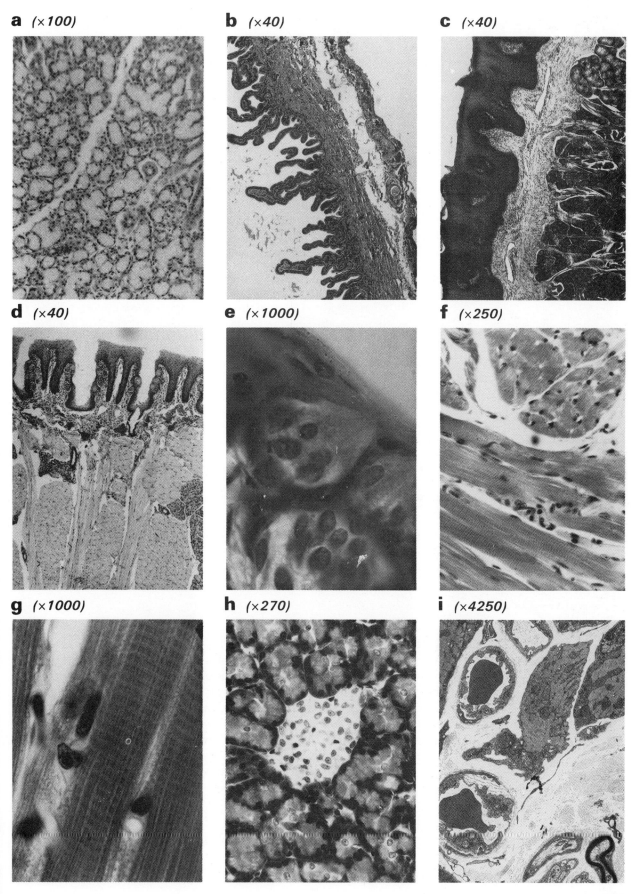

3 Counting

Often in biology, the observations that are made are not descriptive, but are in the form of numerical data. The observations are given a numerical value, that is, they are quantitative data. When the variates consist of whole numbers, the data are described as **discrete**.

Once numbers have been attributed to observations, there is then the opportunity of examining them in a mathematical way, so that they can be made more meaningful. Because of the variable nature of biological material, the use of quantitative data is very important, since then graphical and statistical analyses can be employed – see Volume 1, *Dealing with Data*.

Most things that can be seen clearly can be counted. Using suitable techniques, images can usually be obtained which are within our viewing range. The light microscope can be used to enlarge objects a few hundred times; the electron microscope to enlarge thousands of times. At the other end of the scale are the techniques of aerial photography and satellite imagery. In this section images with a wide variety of scales are used.

3.1 THE PRINCIPLES OF COUNTING

Counting is a skill that must be learned and practised before you become proficient at it. It is very easy to get mixed up or lost while counting objects, whether they are sheep in a field, bacterial colonies on an agar plate or yeast cells under the microscope.

3.1.1 Aids to counting

**Figure 3.1
Tally counter**

The examples and exercises in this section involve the use of photographs and drawings. You can cover the plates with an acetate sheet, and using a washable felt-tip pen, mark each object as you count it, so that you do not miss any out nor count any one twice. Using this method you can also stop and start as you wish. You cannot do this when using a microscope unless you have photomicrograph facilities, but you can count directly in small numbers at a time, subdividing the field of view in some way.

There are a number of ways in which the counts can be recorded:

(i) Using a **tally counter**
     This is a small hand-held mechanical device which is operated by pressing a knob down once for each object counted. It records the number of times it is pressed.

Figure 3.2
Colony counter

(ii) Using a *tally chart*

The method of construction and the use of such a chart is given in Volume 1, page 10 and Work Sheet 1. As each object is counted and marked, a stroke is marked in the chart. The strokes are made up into groups of five. This method of counting is called *gate scoring*. The total frequency is determined at the end.

(iii) Using *specialised counters*

For counting bacterial colonies on agar plates, a specialised counter can be used. This consists of an illuminated stage, a magnifying lens and a special pointer which completes an electrical circuit when it is pressed against the plate. The number of colonies is recorded and displayed.

A number of highly specialised and expensive automatic counters have been designed for counting in particular situations. These are really only practical in research, industry and medicine.

3.1.2 Sampling

Biologists often make general statements from particular observations, however it is important to know the difference between populations and samples. A *population* includes all members of any group which is specified by the researcher. Populations do not necessarily consist of intact organisms; one can deal with populations of parts of organisms.

It is usually not practicable to count all the representatives of a population since this would require far too much effort. The answer lies in taking a *sample* from the population. It is vitally important that the individuals or objects in the sample are fair representatives of the population from which they were drawn. All individuals or elements within the population should have an equal chance of being selected for the sample. Unless the sample is truly representative of the whole population, the effort is wasted. If the sample is biased, then any conclusions can only apply to that sample, and cannot be applied to the whole population. The best method is to take as many samples from the populations as time and effort allows. All samples must be *random*.

Many of the examples and exercises which follow do not indicate that replicates and random samples have been taken, but you should assume that the material is representative of the populations in question. In some exercises the opportunity does arise for you to take your own random samples, for example in 'Sampling weeds in a field', page 116.

3.1.3 The haemocytometer

A haemocytometer is a special slide that can contain a known volume of a liquid, so allowing the total number of cells in the liquid to be determined. Although designed for counting blood cells, the haemocytometer can be used to count any cells that are of a similar size, for example yeasts, single-celled algae and immobile protozoa. It is not suitable for counting smaller organisms such as bacteria; special chambers which contain smaller volumes of liquid have been developed for this, namely the Petroff–Hausser and Halber counting chambers. The preparation and counting procedures for all these chambers are fundamentally the same.

3.2 LEARNING HOW TO COUNT

The following sections give a few examples of how to approach the problem of counting quickly and easily. If you are methodical in your procedure then you should be able to develop the skill with practice.

3.2.1 Counting sheep

Knowledge and skills required	Reference
Use of tally counter or tally chart	3.1.1
χ^2 significance test	Vol. 1 p. 96

Farmers often have to count the numbers of animals in their cattle or sheep herds to make sure that no animal has strayed or become ill. How do they do this so quickly and accurately? Firstly they wait until the animals are feeding quietly and not moving about. They then count them in small groups and add together the totals of the groups to give the grand total.

Look at Plate 3.1(a) and (b) which are of sheep in two fields on a farm.

An acetate sheet was placed over Plate 3.1(a) and obvious groups of sheep circled with a washable felt-tip pen. The numbers of ewes and lambs within each group were then counted, dotting each ewe and lamb as it was counted, and recording the numbers using a tally counter or a tally chart. The total number of ewes and lambs in each group was written in each circle, the final total being the sum of all the groups.

1 Do the same for Plate 3.1(b).

The total number of ewes in Plate 3.1(a) is 43, and the total number of lambs is 50.

2 What is the total number of ewes in Plate 3.1(b)?

3 What is the total number of lambs in Plate 3.1(b)?

4 What is the mean number of lambs per ewe in Plate 3.1(a)?

5 What is the mean number of lambs per ewe in Plate 3.1(b)?

6 Carry out a χ^2 test to find out if there is any significant difference between the numbers of ewes and lambs in flocks (a) and (b).

3.2.2 Eggs in the trematode *Microphallus pygmaeus*

Knowledge and skills required	Reference
Structure and life history of the trematode	Suitable text
Use of tally counter or tally chart	3.1.1
χ^2 test of homogeneity	Vol. 1 p. 102
t significance test	Vol. 1 p. 89

Microphallus pygmaeus is a trematode which is found in the digestive tracts of herring gulls and rock pipits. The larval stages occur in the intertidal gastropod mollusc *Littorina saxatilis*.

PLATE 3.1

a

b

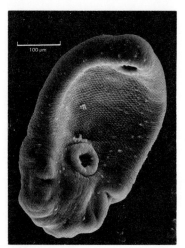

Figure 3.3
SEM Microphallus pygmaeus

In an investigation, specimens of *L. saxatilis* were collected from an island bird sanctuary where large numbers of herring gulls nested each year. In the laboratory the molluscs were dissected, and sporocysts and metacercariae of *M. pygmaeus* were collected. Metacercariae were administered to one-day-old chicks. After six days, individual adult trematodes were obtained and examined under the microscope. The number of eggs that each contained was counted.

Plate 3.2 shows drawings taken from light micrographs of two adult *M. pygmaeus* from the intestines of six-day-old chicks. We want to find out if there is any significant difference between the numbers of eggs found in each of the two trematodes.

1 Why do you think that this trematode was given such unusual generic and specific names?

2 Count the eggs inside trematodes I and II by placing an acetate sheet over the plate. To keep the acetate sheet from moving about make sure that the plate and overlying sheet are put in a file using the locating holes. Using a washable felt-tip pen, put a dot on each egg as you count it. There are two useful ways of recording the egg counts:

(i) For the trematodes, make up a frequency table like Table A below and gate score as you count.

Table A

Trematode	Numbers of eggs										
	5	10	15	20	25	30	35	40	45	50	etc.
I											
II											

(ii) Use a tally counter, pressing it each time you mark an egg so that it automatically counts for you.

If you are not given a tally counter use method (i); if you are given a counter use method (ii).

Using either of these techniques, you can stop at any time during the procedure and still maintain accuracy of counting.

3 Make a note of the total number of eggs that you found in trematode I and the total number of eggs in trematode II.

4 Make a copy of Table B:

Method (i)

Table B

Student	Numbers of eggs	
	Trematode I	Trematode II
1		
2		

etc.

Repeat the table for Method (ii)

PLATE 3.2

a

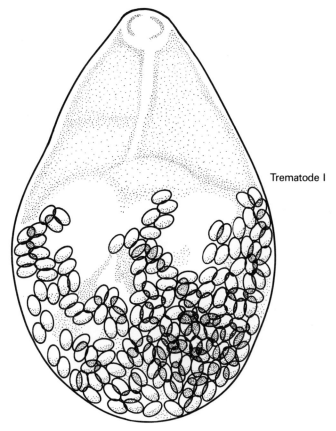

Trematode I

b

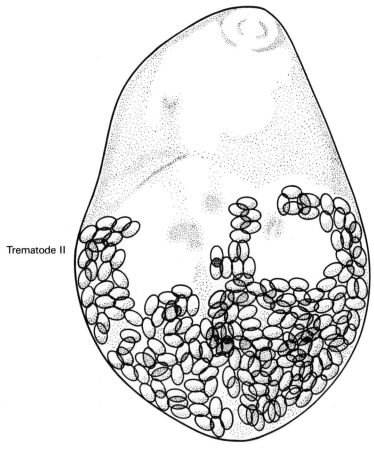

Trematode II

5 Using the χ^2 test of homogeneity, determine if these results are consistent enough for Methods (i) and (ii) separately and together. This should identify any students whose counts are inconsistent. It should also show if there are any discrepancies between the two methods used.

6 Make a copy of Table C. Calculate the mean and variance for each of the boxes shown.

Table C

	Trematode I	Trematode II
Method (i)		
Method (ii)		

7 Using the *t* test find out if there is any significant difference in the counts between
(i) the two methods of counting
(ii) the two trematodes.

3.2.3 *Drosophila:* male and female

Knowledge and skills required	Reference
Male and female *Drosophila*	2.2.4
Tally counting	3.1.1
χ^2 goodness-of-fit test	Vol. 1 p. 99

Much of Mendel's success in unravelling the intricacies and inconsistencies of the results of his breeding experiments lay in the fact that (initially at least) he only dealt with one trait or characteristic at a time.

Carefully examine the group of flies shown on Plate 3.3. It contains 48 flies taken at random from the second generation of flies in a breeding experiment. It consists of a complex mixture of flies, so that it seems to be difficult to make any sense out of it. To overcome this problem, do as Mendel did and count just one characteristic at a time.

From Section 2.2.4 (page 64), you should be able to identify the males and females.

1 Copy out the following table:

Males						Females					
5	10	15	20	25	Total	5	10	15	20	25	Total

2 Gate score (see Volume 1 page 10) the flies in the table, as you identify them. You will find it useful to cover the plate with an acetate sheet and, using washable coloured pens, tick off each fly as you record it in the table.

According to Mendelian theory the ratio of males to females should be 1 : 1.

3 Using the χ^2 goodness-of-fit test (see Volume 1, page 99) determine if your numbers of males and females agree with the 1 : 1 ratio.

PLATE 3.3

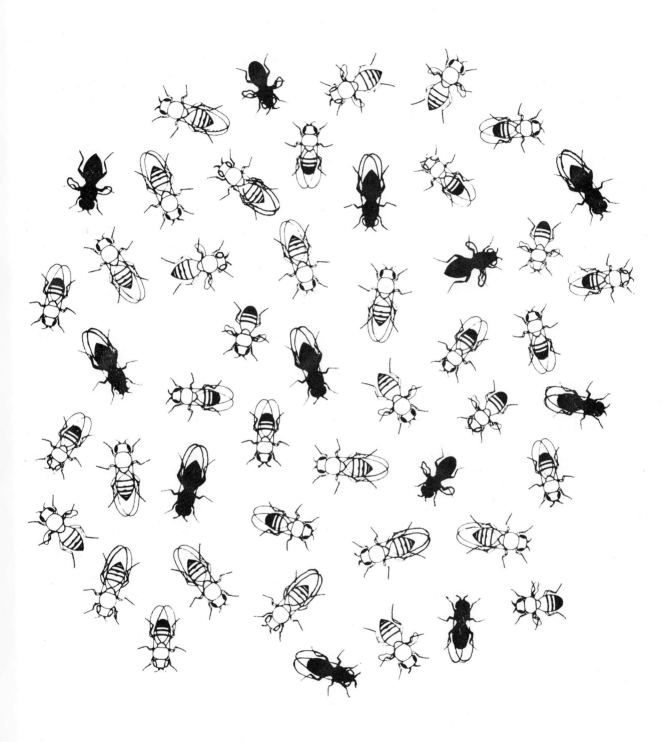

3.2.4 Haemocytometry

The haemocytometer slide (see Section 3.1.3) usually comes in a kit form, with two small pipettes, rubber tubing and mouthpieces, and several special thick cover glasses.

The slide itself is about five times thicker than an ordinary microscope slide, with bevelled edges, and two or three deep grooves cut across the central portion. A slide possessing two parallel grooves has only one chamber, while one with three grooves forming an 'H' shape has two chambers on the same slide (see Figure 3.7) – this allows two counts from the same sample without the need for reassembly of the slide.

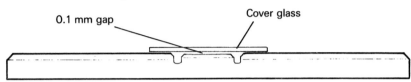

Figure 3.4
Slide viewed from the side

If a cover glass is placed over the grooves and the slide is observed from the side, a small space can be seen below the cover glass between the grooves. When the cover glass is positioned correctly, this gap measures exactly 0.1 mm.

Looking at the slide from above, the area between the grooves is lightly silvered, and by reflecting a light off this surface a tiny cross can be seen etched on the surface of each chamber (see Figure 3.5(a)). When the slide is observed under a microscope at a magnification of about ×40, the cross appears (see Figure 3.5(b)). Nine large squares are visible, each of which measures 1 mm × 1 mm. Each of the large squares is further divided into either rectangles or squares.

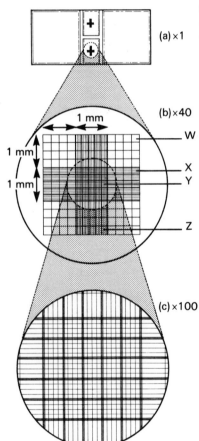

Figure 3.5
Slide viewed at increasing magnifications
a ×1
b ×40
c ×100

1 Referring to Figure 3.5, what are the dimensions of the areas labelled W, X, Y and Z?

For most counting procedures, only the central large square is used. This almost fills the field of view in a microscope when a total magnification of ×100 is used (see Figure 3.5(c)).

The 25 type y squares can now be seen to be bordered by triple lines and to be each further subdivided into 16 smaller squares.

2 What area is marked out by each of these triple-lined type y squares?

3 When the cover glass is correctly in position, what volume of liquid would be found over a y-type square? What fraction of a cubic millimetre is this volume?

Therefore, if the mean number of cells overlying one of these triple-lined squares is determined, the density of cells per cubic millimetre can be calculated by simple multiplication.

4 If x is the mean number of cells per triple-lined square, how many cells are there per cubic millimetre and per cubic centimetre?

When using the haemocytometer several precautions must be taken to ensure that the estimate is as accurate as possible. The procedure involves five basic steps:

1. Dilution of the suspension.
2. Preparation of the slide.
3. Filling the slide.

4. Counting the cells.

5. Computation of the density.

1.   Most cell suspensions contain too many cells to allow accurate counting so the original sample must be diluted by a known number of times. The special pipettes provided in each haemocytometer kit are for use with blood cells, diluting the cell suspension by ×100 with saline solution. Standard serial dilution procedures can be used for other cell suspensions. Initially trial and error must be used to determine a dilution which gives a reasonably even spread of cells across the haemocytometer grid with a minimum of clumping.

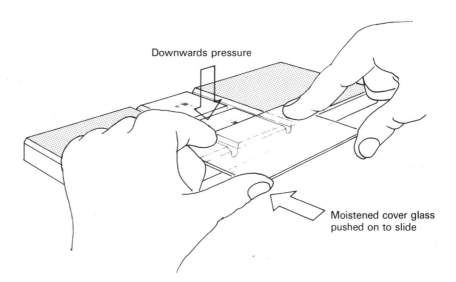

Downwards pressure

Moistened cover glass
pushed on to slide

Figure 3.6
Putting the cover glass on the haemocytometer

2.   Both the slide and the cover glass should be thoroughly cleaned so that any cells that are adhering to them from previous counts are removed. To prepare the chamber it is crucial that the cover glass is positioned correctly. This is done by breathing on the underside of the cover glass and sliding it horizontally on to the top of the slide, pressing down with the index fingers while pushing with the thumbs (see Figure 3.6).

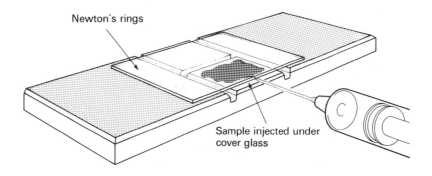

Newton's rings

Sample injected under
cover glass

Figure 3.7
Slide showing Newton's rings and adding the sample

When the cover glass is correctly in place, six rainbow patterns (Newton's rings) should be visible along the two edges of the cover glass where it is supported by the slide (see Figure 3.7). The depth of the chamber is now fixed at 0.1 mm.

3.   A syringe is used to take a sample of the diluted solution which has been well shaken to ensure an even distribution of the cells. A portion of the sample is then carefully injected under the cover glass of the haemocytometer. Only sufficient suspension to cover the silvered front part of the

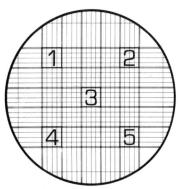

Figure 3.8
Five triple-lined squares
of the haemocytometer

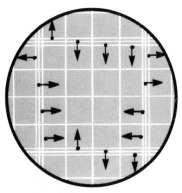

Figure 3.9
The north and west rule

chamber should be used, as any which passes into the grooves is likely to be mainly liquid and such a sample would give an overestimate of the population density. A period of at least five minutes is allowed to elapse before counting is started – this allows the cells to settle on to the grid and avoids having to change focus during counting.

4.   Counting can be carried out on all 25 of the triple-lined squares but this would require a considerable amount of time. As is usual, a sample is taken.

The simplest method is to count the cells in five of the triple-lined squares as shown in Figure 3.8. Counting is normally carried out at magnification ×400 when each of the triple-lined squares fills the field of view.

Plate 3.4 shows one triple-lined square of a haemocytometer slide viewed at a total magnification of ×720. The chamber contains a sample of yeast cells and illustrates the two main problems that arise:

1.   The cells should be evenly spread out. A slide with too much clumping of the cells should be discarded and reassembled with another sample.

2.   Some cells lie on the boundary, that is on the triple lines between adjoining squares. To ensure that such cells are only counted once the conventional method is to count those cells which touch the central line on the **north** and **west** sides of the square, while ignoring those on the south and east.

5 How many yeast cells are there in this square? Place an acetate sheet over Plate 3.4 and accurately rule the north and west boundaries using a washable pen for reference, in case the acetate moves during counting. Count each of the small squares in turn, starting at the top left-hand, and cross out each cell as you count it by marking its position with the felt-tip pen.

Let us assume that the square you have just counted contained the mean number of cells for all the triple-ruled squares.

6 How many cells must have been present in each cubic millimetre of cell suspension?

7 The dilution factor of the sample counted was $\times 10^{-3}$. What was the density of cells in the original suspension?

3.3 APPLICATIONS OF COUNTING

Now that you have developed the skills required for quick and accurate counting, let us look at some applications of this in the form of exercises. Remember what you have already learned, and use whatever techniques you think are most suitable for the problem being studied.

PLATE 3.4

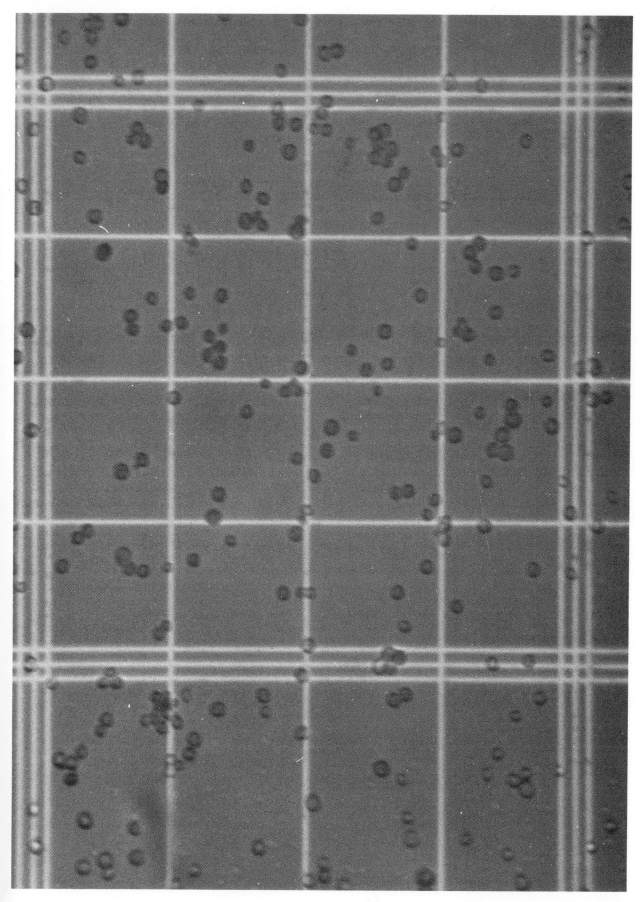

3.3.1 Sampling weeds in a field

Examine Plate 3.5, which is an aerial photograph of a field. In the field are large numbers of a grassland weed called common **ragwort** (*Senecio jacobaea*), which show up as light dots. The farmer needs to know the total number of weeds in this field so that he can decide on how to remove them. Ragwort is poisonous to livestock because it contains alkaloids which affect liver function. Depending on the density of the ragwort, the farmer will either use a biological method of control on the land by grazing sheep (which are more tolerant to the weed and eat it), or a chemical method by spraying weedkiller. It is too time-consuming to count all the weeds in the field, so random samples are taken.

1 Why must the data be collected randomly?

2 What does the term random mean?

How many quadrats are selected as samples? This will depend on a number of factors which have been discussed in Section 3.1.

3 Decide how many quadrats you are going to count.

Various methods have been used to obtain random samples. We will try two of them and compare the results from each method.

Method 1

4 Cut out small acetate squares as instructed (AS1). The number you need is equal to the number of quadrats you decided upon. Now simply scatter them on to the photograph of the field. This is analogous to walking through the field and throwing down quadrat frames.

5 Count the number of ragwort plants in each of the squares, recording your data in a suitable table.

6 Determine the mean number of ragwort plants per 'quadrat', and the standard deviation.

7 Determine the area of the field.

8 From the values you determined in 6 and 7, calculate the total number of ragwort plants in the field.

Method 2

9 Place over the field an acetate sheet (AS2) which has been marked out in squares.

10 Starting as you would with a normal graph, label the lines of the grid, starting in the lower left-hand corner with 00, 01, 02, 03, 04 . . .

11 Obtain a set of random numbers using a computer, or alternatively use a table of random numbers. The total number of pairs you select should be equal to the number of quadrats that you wish to count.

12 Use each pair of numbers you have selected as coordinates to identify a square on the grid that you have drawn; the first number will mark the position on the x (horizontal) axis and the second number the position on the y (vertical) axis. Where the two lines cross is the position of the lower left-hand corner of the sample quadrat.

PLATE 3.5

(i) If the square identified has already been sampled, reject it and repeat the selection process with the next pair of numbers.

  (ii) If the full area of the square identified is not occupied by the field, then reject it and repeat the selection process with the next pair of numbers.

13 Count the number of ragwort plants in each of the sample quadrats identified. As with the haemocytometer (Section 3.2.4), it is necessary that a standard counting procedure should be used to avoid counting the same plant twice. How was this carried out in the haemocytometer?

14 Record all the information about each of the quadrats in a suitable table.

15 From the data, calculate
  (i) the mean density of weeds per quadrat
  (ii) the standard deviation of the sample
  (iii) the total number of ragwort plants in the field.

16 Test your results for each method by carrying out a t test of significance. Was there any difference in the results you obtained by the two methods? Were any differences found due to chance?

17 Which method is the most accurate?

3.3.2 Maize cobs: counts and predicted ratios

Knowledge and skills required	Reference
Monohybrid and dihybrid Mendelian inheritance	Suitable text
Tally counting	3.1.1
χ^2 goodness-of-fit test	Vol.1 p. 99

Maize (Indian corn) occurs in a number of varieties, some of which have white seeds while others have yellow, purple or red seeds. A number of different maize cobs (A–D) are shown on Plates 3.6 and 3.7. Since these are black and white photographs, the colours show up as black, white or grey.

The expected ratios for cobs A–D are:

  A 3 purple : 1 yellow. The purple seeds show up as black and the yellow as white.

  B 3 starch : 1 sweet. The starchy seeds are white and rounded, the sweet seeds grey and wrinkled.

  C 13 yellow : 3 purple. This is a dihybrid cross showing segregation for two colour genes, one complementary and the other an inhibitor.

  D 9 purple : 3 red : 4 white. Purple shows up as black and red as grey. This is a dihybrid cross where the homozygous recessive condition of one of the genes suppresses colour production.

1 Make up a frequency table for each of the cobs.

2 Gate score the seed types found in each of the cobs, including the total numbers. Only count the seeds on rows that you can see clearly.

3 Using the appropriate significance test, find out if the observed numbers agree with the expected ratios.

4 Try and work out how the ratios given above are arrived at.

PLATE 3.6

Cob A

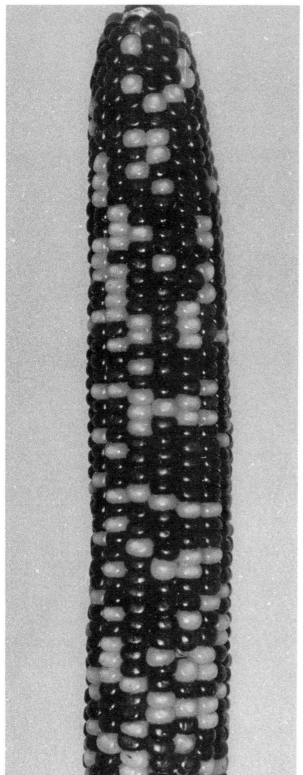

Cob B

PLATE 3.7

Cob C

Cob D

3.3.3 Population growth in yeast

Knowledge and skills required	Reference
Use of the haemocytometer	3.2.4
Life cycle of yeast	Suitable text
Construction of log–linear graph	Vol. 1 p. 25
Log–linear equation	Vol. 1 p. 53
Confidence limits	Vol. 1 p. 84
Population dynamics	Vol. 1 p. 25

You found out in Section 3.2.4 (page 112) that yeasts are very suitable organisms to count using a haemocytometer slide. They can also be used as a model organism when we want to find out how populations increase or decrease.

In an experiment, a small quantity of a sugary culture medium was inoculated with a standard number of yeast cells. After predetermined periods of time, random samples were removed from the culture and placed on a haemocytometer slide.

Plates 3.8 and 3.9, on the following pages, show drawings of three replicate fields of view from the haemocytometer slide at each sample time, viewed under the high power of the microscope.

1 Using the techniques explained in Section 3.2.4 (page 112), count the number of yeast cells in each of the fields of view for the four hour sample. Calculate the mean value and the 95 per cent confidence limits.

2 Why is it not necessary to determine the total number of yeast cells in the culture flask?

3 Repeat the counts and calculations for the other four samples shown. Organise the data into a suitable table.

4 Illustrate the changes in the yeast population by drawing a suitable graph.

5 Describe the changes which have taken place in the yeast population during the experiment.

6 Suggest reasons for the changes.

7 Use a graphical or mathematical method to estimate the size of the yeast population   (i) at time 0   (ii) at time 32 hours.

3.3.4 *Sordaria* asci: tetrad analysis

Knowledge and skills required	Reference
Ascomycete fungi	Suitable texts
Meiosis	Suitable texts
Cross-over values and chromosome mapping	Suitable texts
χ^2 goodness-of-fit test	Vol. 1 p. 99
Tally counting	3.1.1

The ascomycete fungi have become almost as famous as *Drosophila* as a research tool in genetics. They are the largest group within the fungi and are characterised by the production of an elongated spore sac called an **ascus** (plural: asci).

1 Suggest reasons why this organism has been found to be so suitable for genetics research.

PLATE 3.8

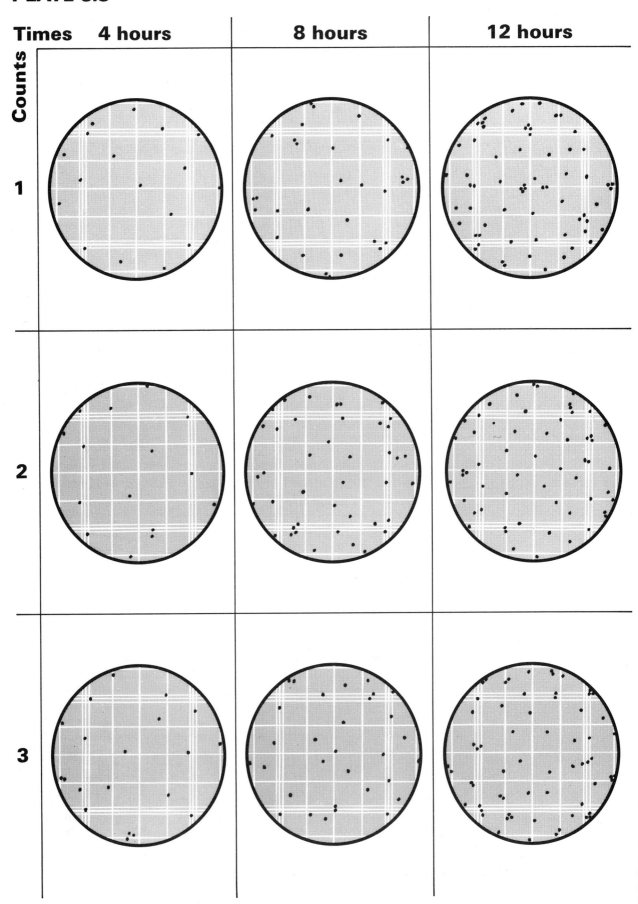

Times	4 hours	8 hours	12 hours
Counts			
1			
2			
3			

PLATE 3.9

Times	16 hours	20 hours

Counts

1

2

3

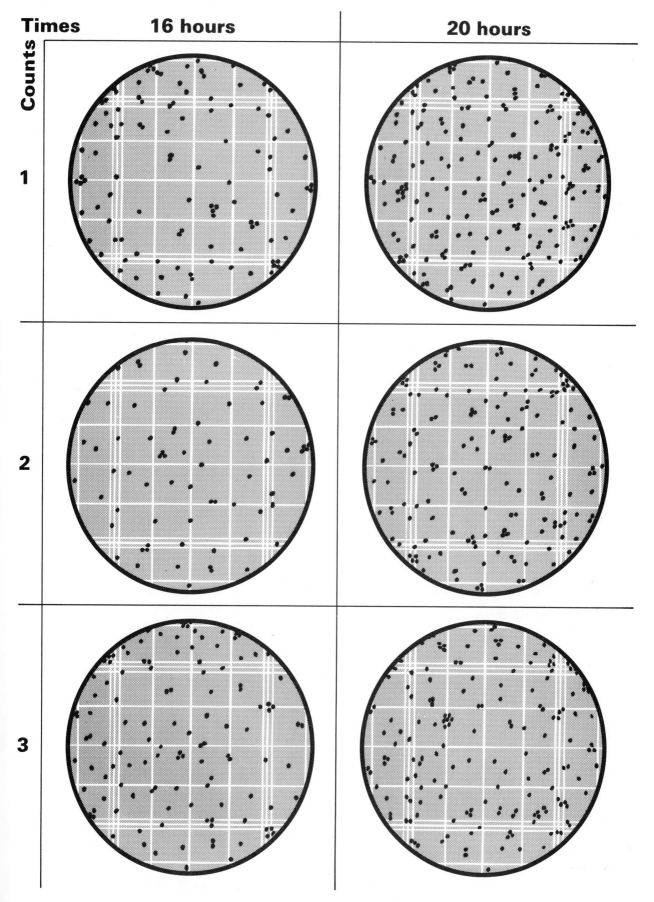

Two main species of ascomycetes have been used in research – *Neurospora crassa* and *Sordaria fimicola*. Lindegren (1932) carried out tetrad analysis as in the following exercise but on *Neurospora*, the same species that Beadle and Tatum used to test their one gene–one enzyme hypothesis. *Sordaria* is much easier to culture than *Neurospora* and so is more often used for such work.

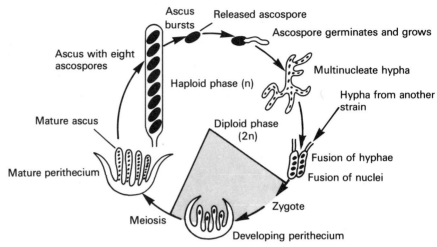

Figure 3.10
Life cycle of Sordaria fimicola

The ascomycetes are haploid for most of their life cycle but two different strains can come together during a sexual phase. This produces fruiting bodies called **perithecia**, inside which develop up to 300 asci. Each zygote in a developing ascus divides first of all by meiosis and then by mitosis to produce eight ascospores.

The most important point about these organisms is that the pod-shaped ascus is so narrow that the developing spores are kept in a line as they develop. They cannot slide past each other, so that the order of the spores in the ascus mirrors what has happened to the chromosomes and chromatids during meiosis.

The ascospores are normally black, but there is a mutant strain which has white spores.

Black- and white-spored strains are inoculated on to a corn-meal agar gel in a Petri dish. The hyphae grow outwards and where they meet in the mid-line, a row of dark peritheciae can be seen (Figure 3.12).

a Black-spored Sordaria

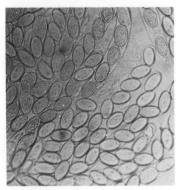

b White-spored mutant Sordaria

Figure 3.11

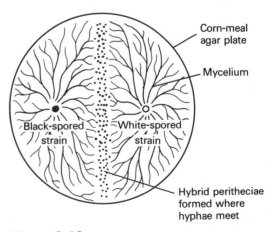

Figure 3.12
Growth of white and black Sordaria *strains*

One of these fruiting bodies is placed on a microscope slide with a drop of water, and tapped gently under a cover slip so that the asci spread out as in Plate 3.10. Each ascus contains eight ascospores because meiosis results in the zygote nucleus producing four nuclei, and this is followed by one mitotic division giving eight nuclei and thus eight haploid ascospores.

From Plate 3.10, you can also see that each ascus is a hybrid, containing four black and four white spores, although the arrangements can be different.

2 Identify and draw the different arrangements of the eight ascospores.

How are these linear arrangements of spores arrived at? The process of meiosis involves two divisions (anaphase I and II, when the chromatids separate to the poles of the spindle). Crossing-over occurs during prophase, and prior to anaphase I. Examine Figure 3.13 which shows how the different arrangements come about.

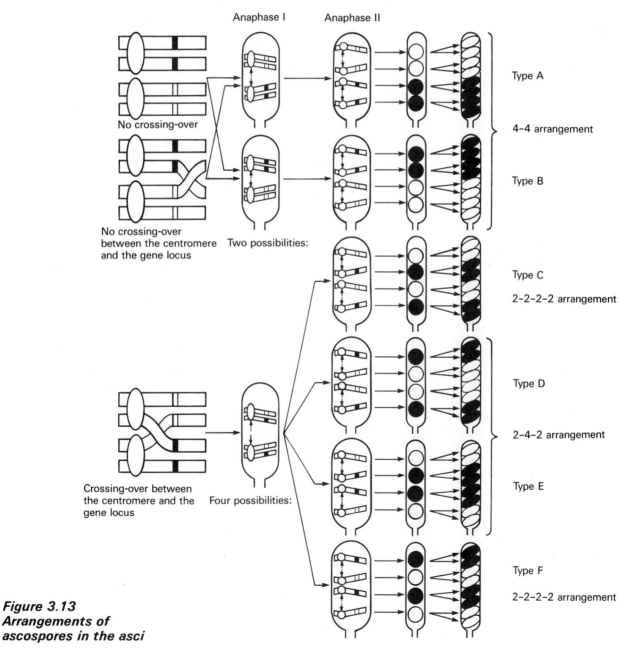

Figure 3.13
Arrangements of ascospores in the asci

PLATE 3.10

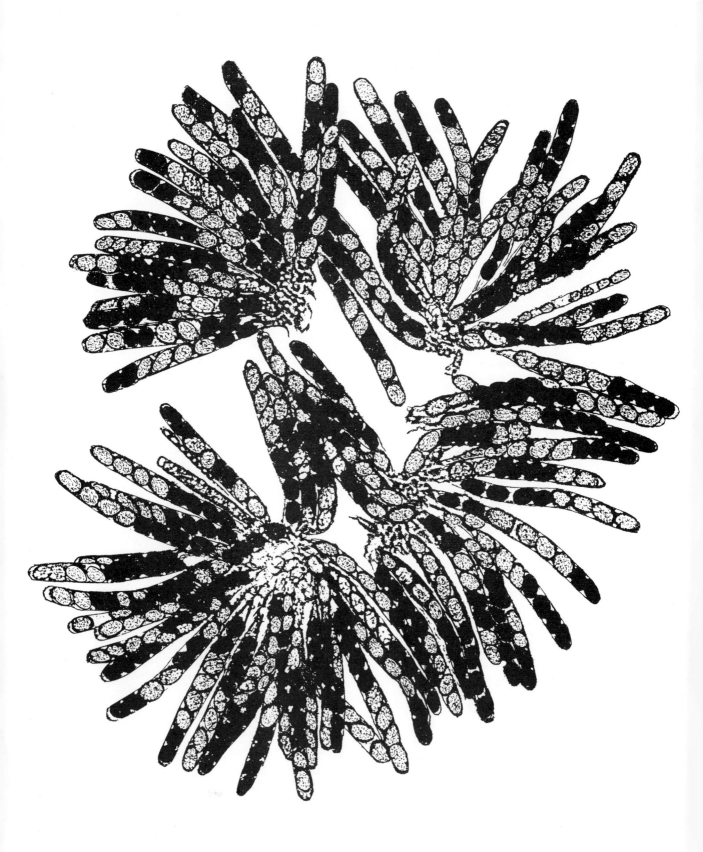

3 Examine Plate 3.10 again, showing the asci in one hybrid perithecium. Count the number of each of the spore arrangements found in the contents of a single hybrid perithecium for types A–F (given in Figure 3.13). Make out and complete a suitable table and enter your data.

4 Do you notice anything about the numbers of each type?

In Figure 3.13 you can see that in asci types A and B, no crossing-over occurs between the centromere and the gene locus (position of the gene) for spore colour. This means that there is an equal chance that the paired chromatids will move to either pole during anaphase I. There should therefore be equal numbers of type A and type B asci.

5 Carry out a χ^2 test on your numbers of asci of types A and B to find out if there is any significant difference between them.

Asci types C, D, E and F occur when crossing-over takes place between the centromeres and the gene locus. From Figure 3.13 we can see that there is an equal chance of any of the four arrangements occurring, since the chromatids can move to either pole during anaphase II. There should therefore be equal numbers of asci of types C, D, E and F.

6 Carry out a χ^2 test on the numbers of asci of types C, D, E and F to determine if there is any significant difference between them.

Since the numbers of cross-overs (C, D, E and F asci) has been determined, we can find out what their frequency is, and this can be used as a unit of map-distance between the centromere and the gene.

The percentage cross-over value (COV) is given as

$$\frac{\frac{1}{2}(\text{the total number of cross-over asci})}{\text{the total number of asci}} \times 100$$

Note that the number of cross-over asci is halved, because half of the chromatids are not involved in crossing-over.

7 What is the COV for these asci?

8 What is the map-distance from the centromere to the gene locus for spore colour?

By using the COVs for many genes it is possible to determine the distance between them, and thus construct chromosome maps. This has been done for some species, such as *Neurospora* and *Drosophila*.

9 Search suitable texts to find examples of chromosome maps of *Neurospora* and *Drosophila*.

3.3.5 Human chromosomes: abnormalities

Knowledge and skills required	Reference
Stages in mitosis	Suitable text
Tally counting	3.1.1

It was in 1956 that Tjio and Levan first clearly demonstrated that there are 23 pairs ($2 \times 23 = 46$) chromosomes in man: 22 pairs of them being **autosomes** and one pair the **sex chromosomes**.

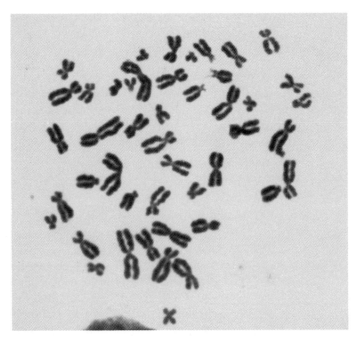

Figure 3.14
Human chromosomes at mitotic metaphase

Figure 3.14 shows a set of human chromosomes at metaphase. You can see that each chromosome consists of two chromatids that remain attached to the common centromere – these will separate at anaphase.

It was soon realised that studying human chromosomes in detail could tell us a considerable amount about the incidence of human genetic disorders, and techniques were developed to make arranging and counting the chromosomes more straightforward. This method is known as **karyotyping**, and involves transferring cells to a nutrient solution containing phytohaemagglutinin

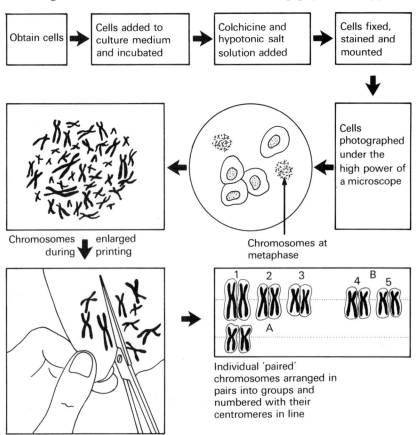

Figure 3.15
Making a karyotype

which stimulates them to divide by mitosis. After a few days the cells are checked to make sure that they are dividing and the drug colchicine is added to the culture – this prevents the formation of spindle fibres.

1 What benefit would adding colchicine have?

A very weak (hypotonic) salt solution is then added to the cell culture.

2 What is the reason for doing this?

The cells are then examined under the high power of a light microscope and the chromosomes photographed and enlarged during printing to give a print like Figure 3.14. The individual chromosomes, each consisting of two chromatids, are then cut out from the enlarged print using scissors, and the 22 autosomes are arranged (in homologous pairs) into seven groups, A–G. The XX or XY (sex) chromosomes are in separate groups.

This process of karyotype analysis is an important test for detecting possible genetically abnormal offspring, and for the recognition of potential genetic congenital malformations resulting from chromosome abnormalities. It was found, however, that classifying the chromosomes was difficult; more information about the individual chromosomes was required to tell the difference between them.

The chromosomes are subjected to a particular form of treatment. Figure 3.16 is a photograph of a treated set of chromosomes.

3 Examine the metaphase chromosomes in Figure 3.16, and compare them with those shown in Figure 3.14. What difference do you notice between them?

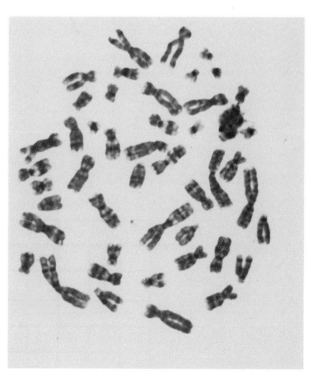

Figure 3.16
Treated set of human
metaphase chromosomes

A complete set of simplified diagrammatic chromosomes set out as a karyotype is shown in Figure 3.17 overleaf.

Let us examine a set of **normal** human chromosomes that have been treated as indicated above with colchicine and trypsin etc. – see Plate 3.11(a).

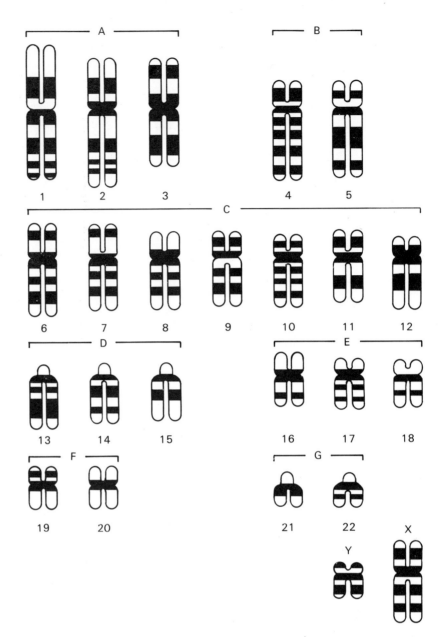

Figure 3.17
Stylised human karyotype

4 Count the double chromosomes in Plate 3.11(a) and make a note of the number

5 Copy out the following table:

A			B		C						
1	2	3	4	5	6	7	8	9	10	11	12

D			E			F		G			
13	14	15	16	17	18	19	20	21	22	X	Y

6 Cover Plate 3.11(a) with an acetate sheet and tick off the chromosomes as you match them with those given in the stylised karyotype in Figure 3.17. Each time you mark a chromosome pair enter a stroke in the appropriate place in the table you have drawn.

PLATE 3.11

a

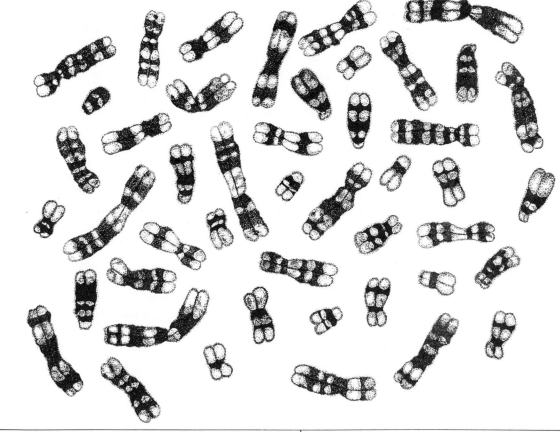

b

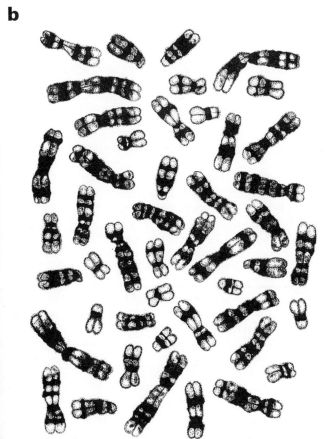

c

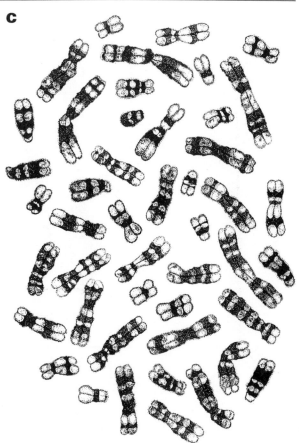

7 Does this set of chromosomes belong to a male or a female?

If a cell comes from a female, it contains two X chromosomes, that is XX. It also contains a small chromatin-positive body lying just inside the nuclear membrane. This is known as a Barr body after Murray Barr who first described it in 1949. The presence or absence of a Barr body can thus be used to determine the sex of a foetus. Smears of cells from the inside of the cheek can be examined for Barr bodies to determine the sex of individuals, for example athletes entered for the Olympic Games.

A lot of research is concentrated on the problem of keeping the risk of birth abnormality to a minimum. Where there is a family history of inherited illness or abnormality, intending parents can seek advice at a genetic-counselling clinic. Here a test known as **amniocentesis** can be carried out very early in a pregnancy to find out if the foetus is normal. This test involves removing a small quantity of the amniotic fluid which surrounds the developing foetus. An ultrasound scan is carried out beforehand to determine the exact position of the foetus and placenta and hence the best place to insert a small needle through the anterior abdominal wall.

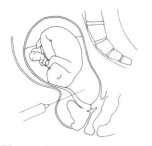

Figure 3.18
The technique of
amniocentesis

With this technique there is a small risk of causing miscarriage, so it is recommended only when there is already a considerable risk of foetal abnormality. More recently, a technique has been developed which involves sucking embryonic cells from the chorionic villi into a narrow tube inserted into the womb via the vagina. Even more useful is the recent development of a blood analysis technique which can indicate whether it is necessary for a mother to have the amniocentesis test or not.

Karyotyping thus makes it possible to detect foetuses that might have abnormal chromosome numbers at an early stage in gestation, when a decision can be taken whether or not to terminate the pregnancy. The technique can also be used on white blood cells to diagnose chromosome abnormalities after birth.

Let us look at some abnormal chromosome counts.

8 Examine Plate 3.11(b), on the previous page. Count the chromosomes carefully. What do you find?

9 Go through the chromosomes carefully, numbering them by using the karyotype given in Figure 3.17.

10 What conclusion can you come to regarding these chromosomes?

This is probably the most common congenital abnormality in man, with a frequency of 1 in 600 births. The condition was first described in detail by Langdon-Down in 1866 and is known as **Down's syndrome**. The patients are of small stature and have mental retardation which can range from mild to severe.

One of the features of Down's syndrome is shown by the following data:

Age of mother	Number of affected offspring per 1000 births	
	Risk of occurrence	Risk in subsequent children
20–30	0.66	2
31–35	1.33	4
36–40	1.66	5
41–45	3.33	10
46 and over	15.00	50

PLATE 3.12

a

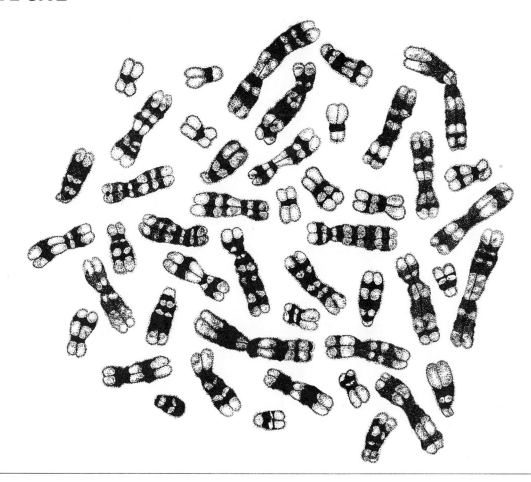

b

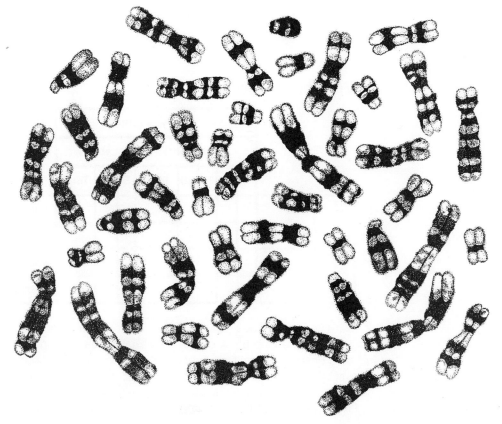

11 Draw a suitable graph to illustrate these data.

12 What conclusions can you come to regarding these data?

13 Suggest as many hypotheses as you can to explain your answers to Question 12.

The chromosome arrangement found in Plate 3.11(b), page 131 is found in about 95 per cent of Down's children. The other 5 per cent, shown in Plate 3.11(c), occurs in children of younger mothers, and is much more likely to run in families.

14 Examine Plate 3.11(b) and compare it with Plate 3.11(c). What differences can you find?

15 Repeat Questions 9 and 10 for Plate 3.12(a) and (b), page 133.

16 What effect do you think the abnormalities had on the individuals concerned in Plate 3.12(a) and (b)?

3.3.6 *Drosophila:* crosses and phenotype counts

Knowledge and skills required	Reference
Ability to recognise male and female and various mutants of *Drosophila*	2.2.4
Gate scoring	Vol. 1 p. 10
Monohybrid cross	Suitable text
Dihybrid cross	Suitable text
χ^2 goodness-of-fit test	Vol. 1 p. 99

You found out how to count male and female *Drosophila* in Section 3.2.3 (p. 110).

1 Copy out the following table:

		Body colour										
		wild (light)					ebony					
		5	10	15	20	Total	5	10	Total			
Wing type	wild (long)									wild (red)		Eye colour
										white		
	vestigial									wild (red)		
										white		

2 Examine Plate 3.3 again which shows a second generation *Drosophila* cross (red eyes show up black on the plate). Taking each fly in turn, score for all three characteristics, for example if it has long wings, a light body and red eyes, then it is scored in the top left-hand cell of the table. Using the gate scoring technique, allocate all the flies to the appropriate cells of your table. Finally determine the total number of flies for each of the eight cells.

Monohybrid cross

In a monohybrid cross, only one pair of contrasting characters is considered at a time, for example long compared to vestigial wings. We can compare the observed numbers in a second generation with those that we would expect according to Mendel's theories. The expected ratio in the second generation (F_2 generation) of such a cross is three dominant to one recessive (the mutant forms in this case are all due to the presence of a recessive gene).

3 Examine the results in the table and write down the total number of flies with long wings and the total number of flies with vestigial wings. Carry out a χ^2 goodness-of-fit test to determine if these numbers agree with a 3 : 1 ratio.

4 Repeat Question 3 but this time compare the numbers of wild type (light-coloured) flies with ebony flies.

5 Repeat the procedure, comparing the number of flies with red eyes with those with white eyes.

Dihybrid cross

Let us now consider two traits together, for example wing type and body colour.

6 Determine from the table the total numbers of types of fly with:
   (i)   long wings and light body
   (ii) long wings and ebony body
   (iii) vestigial wings and light body
   (iv) vestigial wings and ebony body.

According to Mendelian theory the ratio in the F_2 generation of a dihybrid cross is 9 : 3 : 3 : 1.

7 Carry out the χ^2 goodness-of-fit test, to determine if the observed numbers for each of the types above agree with this ratio.

8 Repeat the procedure for wing type and eye colour.

9 Repeat for body colour and eye colour.

Trihybrid cross

Mendel actually went on to consider three traits at once. You could try to find out if the numbers agree with what would be expected.

10 Determine from your table the numbers of flies with the following characteristics:
   (i)   long wings, light body and red eyes
   (ii)   long wings, light body and white eyes
   (iii) vestigial wings, light body and red eyes
   (iv)   vestigial wings, light body and white eyes
   (v)   long wings, ebony body and red eyes
   (vi)   long wings, ebony body and white eyes
   (vii) vestigial wings, ebony body and red eyes
   (viii) vestigial wings, ebony body and white eyes.

11 The ratio in the F_2 generation for a trihybrid cross is 27 : 9 : 9 : 9 : 3 : 3 : 3 : 1. Carry out the χ^2 goodness-of-fit test to find out if the numbers you obtained agree with this ratio. (Usually, to be valid, this test should not be used if any category has less than five flies.)

3.3.7 Mitochondria in the tail of trematode cercariae

Knowledge and skills required	Reference
Life cycle of trematodes	Suitable text
χ^2 test	Vol. 1 p. 96
2×2 contingency table	Vol. 1 p. 104

The digenean trematode *Cryptocotyle lingua* has a complex life cycle which includes a primary molluscan host, an intermediate fish host and a sea-bird as the final host. It has the usual multiplication phase in the mollusc host which ends in the release of very large numbers of **cercariae** into the water. In *C. lingua*, the cercariae do not feed and so have to find the fish host quickly.

The cercariae are **positively phototactic**, that is they swim towards the light.

Figure 3.19 shows the cercariae swimming towards a light source which is at the top of the photograph.

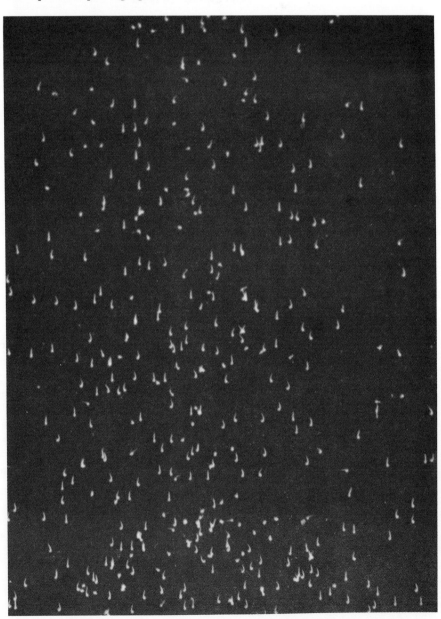

Figure 3.19
Swimming cercariae of
Cryptocotyle lingua

1 How many cercariae do you estimate are shown in this photograph?

2 Find out if you are right.

Cercariae were obtained from the seashore mollusc *Littorina littorea* and prepared for TEM microscopy. Sections were taken (see Figure 3.20) from the base and middle regions of the tails of cercariae which had been released for one hour, and from cercariae which had been allowed to swim around for ninety-six hours.

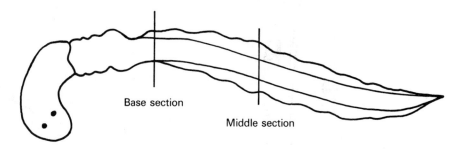

Figure 3.20
Cercaria of Cryptocotyle lingua

Base section

Middle section

Plate 3.13 shows two sections (base and middle) from a newly released cercaria.

Plate 3.14 also shows two sections (base and middle) but from a cercaria which has been released for ninety-six hours.

3 Count the mitochondria in each of the four micrographs.

A tentative hypothesis suggested that there are more mitochondria in the middle region of the tail than in the base, because it was thought that the main propulsive forces are generated in the middle region. A second hypothesis suggested that the mitochondria degenerate over time as the energy reserves become exhausted.

4 Using suitable statistical tests find out if the tentative hypotheses are correct.

5 Examine the micrographs carefully and describe any differences that you can see between the old and the new sections.

PLATE 3.13

a

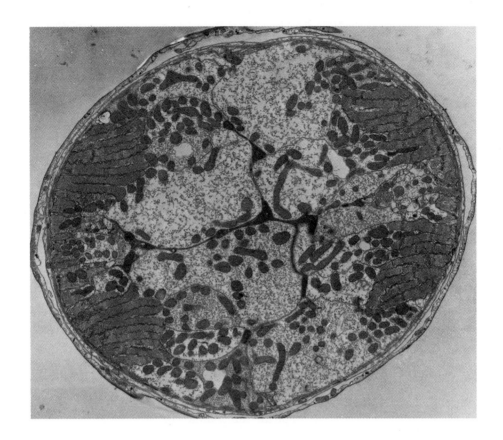

b

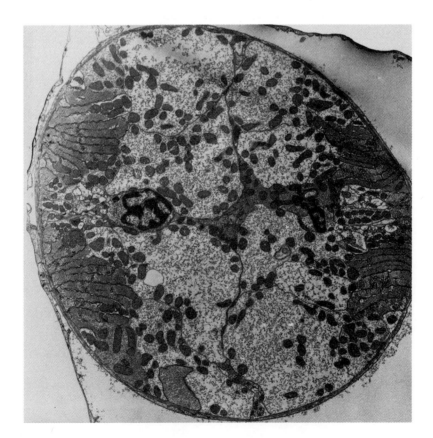

PLATE 3.14

a

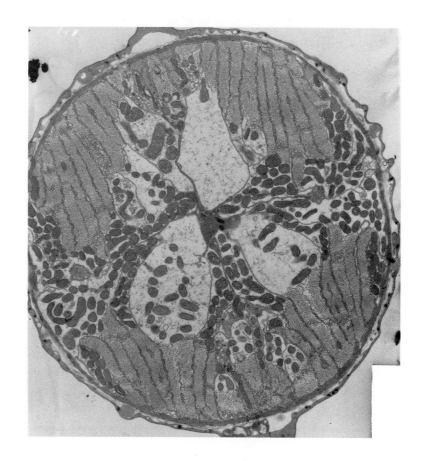

b

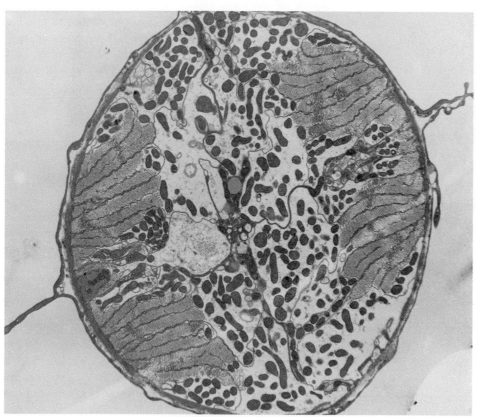

4 Measuring

Measuring is a form of counting in which units are counted instead of items. Measuring is thought of in relation to lengths, widths, area (including surface area), volume, weight and time.

In this section we shall be mainly looking at those things in biology which can be difficult to measure. Biological structures are normally three-dimensional but we usually have only two-dimensional images to work from. Special techniques have been developed to overcome this problem.

Measuring, like counting, produces **quantitative** data which can be dealt with graphically and statistically (see Volume 1, *Dealing with Data*).

4.1 THE PRINCIPLES OF MEASURING

As long as we have the object or an image of the object that we want to measure lying within our normal range of vision, then we can measure it without difficulty. It does not matter whether the image has been obtained with an electron microscope or using aerial photography, the techniques are the same. The main problems arise when we are dealing with three-dimensional objects, and only have two-dimensional images to work with, for example microscope sections.

4.1.1 The unaided eye

For objects that lie within our normal range of vision, everyday measuring equipment can be used, for example rulers, vernier callipers and micrometer screw gauges. These are adequate for measuring the length and width of mollusc shells, leaves and so on. Very small structures can be measured directly from a micrograph.

4.1.2 Micrometry

Micrometry is the biometry of specimens which are too small to be measured accurately with the unaided eye – it is measuring with a microscope. The type of microscope that is used depends on the size of the specimen to be measured and thus how much it must be magnified.

A glass disc with a scale of arbitrary length (often 10 mm) etched on it, is placed inside the eyepiece of the microscope. This **eyepiece graticule** usually contains one hundred divisions, although a wide range of scales is available. When an object is viewed through the microscope, the eyepiece graticule is superimposed on the field of view. See Figure 4.1.

The arbitrary units of the eyepiece scale may be used to compare the dimensions of all specimens observed under the same magnification and with the same microscope or type of microscope. If we want to compare specimens at different magnifications or use different techniques, then the eyepiece scale must be calibrated to the appropriate SI unit, the micrometre (μm).

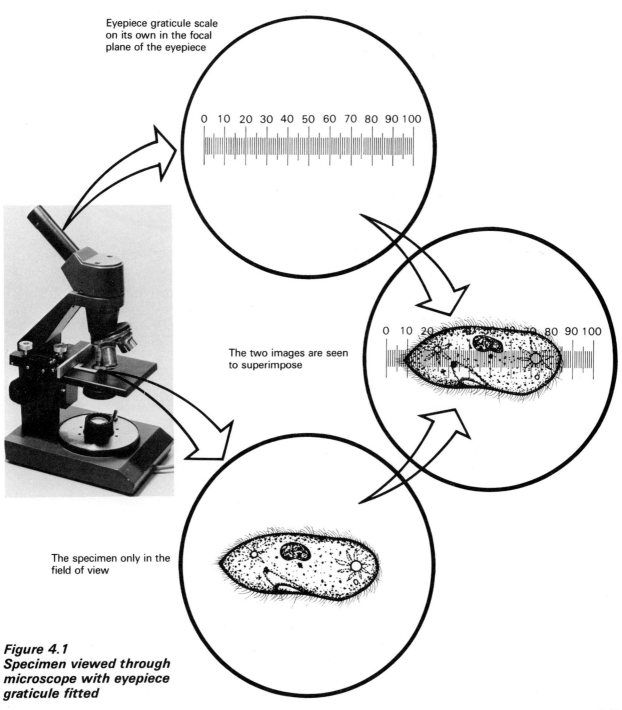

Eyepiece graticule scale on its own in the focal plane of the eyepiece

The two images are seen to superimpose

The specimen only in the field of view

Figure 4.1
Specimen viewed through microscope with eyepiece graticule fitted

To calibrate the eyepiece graticule a scale of known length, called the **stage** or **object micrometer**, is placed on the stage of the microscope. On looking through the microscope and focusing on the stage micrometer, the two scales superimpose as in Figure 4.2:

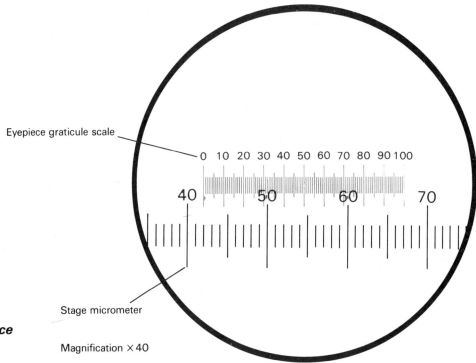

Eyepiece graticule scale

0 10 20 30 40 50 60 70 80 90 100

40   50   60   70

Stage micrometer

Magnification × 40

Figure 4.2
Superimposed eyepiece graticule and stage micrometer scales

The total length of the 100 eyepiece graticule divisions can easily be measured and thus the length of each small division determined. The calibration has to be repeated for each objective lens on the microscope.

1 Why do you think it is necessary to do this?

Although these principles normally apply to monocular microscopes, they can be applied to low-power stereomicroscopy, when objects in the range 0.1–10 mm are being measured.

4.1.3 Photomicrography

Photomicrography has been one of the most important developments since the development of the microscope itself. It can be used over a very wide range of magnifications, from low-power light microscopy to very high-power electron microscopy. At any magnification it is useful to have actual photographs of what has been observed, so that the material may be examined and analysed at any time. Also, other scientists have the opportunity to interpret exactly the same images.

If the resolution is good, for example when using the EM, enlargements may be taken from the negative to give much higher secondary magnifications. Many of the figures and plates in this book are enlarged micrographs.

A lot of basic information about cell structures can be obtained by simply measuring the dimensions of such structures on light or electron micrographs. Measurements are usually made in millimetres.

1 If we wanted to convert our measurements in millimetres into micrometres, what would we multiply our values by?

2 If we wanted to convert our measurements in millimetres into nanometres, what would we multiply our values by?

We can use length measurements from thin sections to calculate the surface areas and volumes of cells and cell components if they have a regular shape. A number of simple formulae can be used:

	Surface area	Volume
Cuboidal	$6L^2$	L^3
Columnar	$4LH + 2L^2$	$L \times L \times H$
Spherical	$4\pi R^2$	$\frac{4}{3}\pi R^3$
Cylindrical	$2\pi R (R + H)$	$\pi R^2 H$
Ellipsoidal	$4\pi ab$	$\frac{4}{3}\pi ab^2$

L = length; H = height; R = radius; a = major axis; b = minor axis

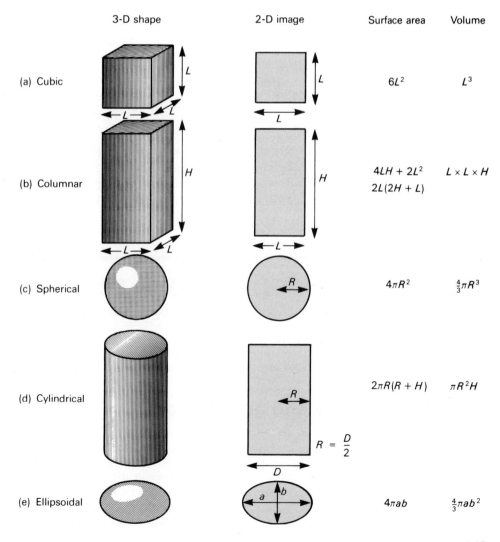

Figure 4.3 Measurement of cells and cell components

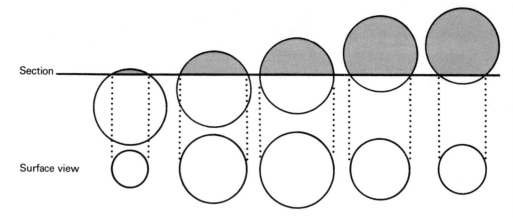

Section ——

Surface view

Figure 4.4
Sections and size

Thin sections will rarely pass directly through the middle of cellular structures so that the sizes of spheres, for example nuclei, vesicles, etc., are difficult to assess accurately (see Figure 4.4). Mean values of sizes can be used as a more accurate measurement.

4.1.4 Measuring cell components by stereology

Stereology is an indirect method of determining values for the surface area and volume of cell components. It is a fair assumption to make that the frequency of cell components in thin sections is related to the amount of them in the whole cell. The problem of determining volumes of cell components is really one of determining surface areas, since ultra-thin sections are always used.

Estimation of the surface areas and volumes of cell components is achieved by placing a grid over the micrograph. Various types of grid can be used depending on what we want to measure; they can be square lattices, point lattices, lines, interrupted line lattices and so on. Details on how these lattices are used are given in Section 4.2.2.

There is a straightforward method for determining the relative volumes of cell components. The micrograph is photocopied on to card, and the various cell components are cut out and weighed. From these weights the percentage contribution of each component to the cell can be calculated.

4.2 LEARNING HOW TO MEASURE

The following section provides you with a variety of photographic material and indicates in each case how the various structures can be measured.

4.2.1 Micrometry and graticules

Knowledge and skills required	Reference
Use of eyepiece graticule	4.1.2
Use of stage micrometer	4.1.2
Drawing linear graphs	Vol. 1 p. 15

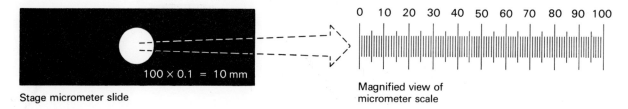

Stage micrometer slide

Magnified view of
micrometer scale

Figure 4.5
Stage micrometer slide
(actual size) and
magnified micrometer
scale

As described in Section 4.1.2 (page 140), a stage micrometer can be used to measure the actual length of the eyepiece graticule scale placed in a microscope. Figure 4.5 shows a typical stage micrometer.

1 What is the total length of the scale in the clear central portion? Use appropriate SI units.

2 How many divisions are there?

3 What is the size of one of the small divisions?

When a microscope is fitted with both an eyepiece graticule and a stage micrometer, an image appears as in Figure 4.2, page 142.

4 What is the total length of the eyepiece scale when measured by the stage micrometer?

5 What is the length of the smallest division of the eyepiece scale in milli-metres; in micrometres?

When the stage micrometer is viewed using different objectives of the micro-scope the images appear as in Figure 4.6, overleaf.

6 Repeat Questions 4 and 5 for magnifications of ×100, ×400 and ×1000, as shown in the diagrams.

7 Copy out the table given below and enter your results.

Total magnification	Total length of 100 divisions of the eyepiece graticule (μm)	Length of each division of the eyepiece graticule (μm)
×40		
×100		
×400		
×1000		

The eyepiece graticule in the microscope has now been calibrated for the various objectives, and the stage micrometer can be removed since it is no longer needed. The dimensions of any very small specimen can now be expressed in micrometres by measuring with the eyepiece graticule scale and converting to micrometres by multiplication.

8 What type of graph line do you think would be produced if the results of such calculations were plotted on a linear grid showing the number of eyepiece divisions against the actual length?

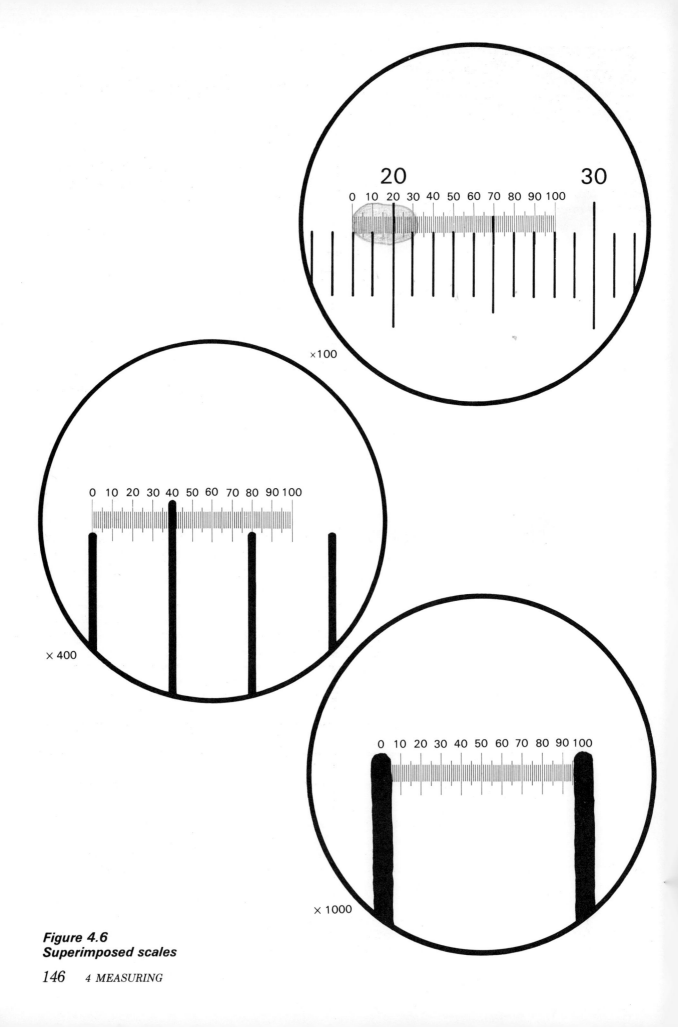

Figure 4.6
Superimposed scales

9 From the table, the actual length of one eyepiece division and of 100 eyepiece divisions are known for each magnification. Use the data for the ×40 objective to plot two points on a linear grid and interpolate to produce a straight line.

Such a graph is known as a **conversion graph** since it can be used to convert one variable into another one.

10 Using appropriate scales on the x- and y-axes, plot lines for each of the magnifications on one grid.

This compound conversion graph can quickly and accurately be used to determine the size of any organism or structure using any of the magnifications available on the microscope.

4.2.2 Stereological techniques

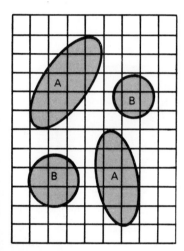

Figure 4.7
Percentage area occupied by two cell components, A and B

Knowledge and skills required	Reference
The cell	2.2.1
Tally counting	3.1.1
Stereology	4.1.4

These recently developed techniques have many applications in diverse fields of research. To find out exactly how they work you can make a number of measurements on each micrograph supplied. Each time, rotate the micrograph by 90° relative to the grid. In research work a wide range of micrographs of the same material would be available.

To assess the areas occupied by two cell components, A and B, we can use a grid made up of squares with a side length of 20 mm for a normal photograph print size of 10 × 8 inches. We can then count the number of squares occupied by each of the components.

In Figure 4.7, the area occupied by A is 18 squares and by B is 10 squares. There are 108 squares in total, so the fraction of the total area occupied by A will be 21/108 = 0.194 or 19.4% and for B it is 12/108 = 0.111 or 11.1%.

We can make the estimation simpler by replacing each square with a central point, as shown in Figure 4.8.

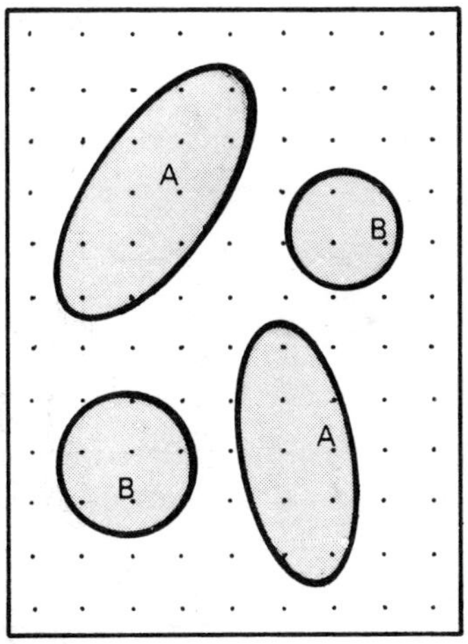

The total number of points on the lattice is 108. Within the boundary of A there are 23 points and within B there are 11 points. The area fraction occupied by A is therefore 23/108 = 0.213 or 21.3% and by B is 11/108 = 0.102 or 10.2%.

These figures are reasonably comparable with the figures found by the first method but the accuracy would obviously be improved by increasing the number of dots. It is a matter of balancing the accuracy required against the time available – the more dots there are, the more time is required. For a 10 × 8 inches print there should be about 100–400 points.

If the components shown in a photograph or micrograph are large, then the lattice with widely spaced points, for example 2 cm apart (the **major** lattice), can be used. For the measurement of smaller components, the lattice with more points, for example spaced at 1 cm apart (the **minor** lattice), should be used.

Figure 4.8
Estimating the percentage area using a point lattice

Counting the number of points over each cell component and over the whole micrograph thus gives an estimate of the surface areas of the components relative to each other and to the whole surface. These surface areas really represent relative volumes since the micrograph represents a thin section.

The areas of the membrane surfaces can also be determined as the surface area per unit volume (in units of $\mu m^2/\mu m^3$), where the value represents the interface surface area. ***It is accepted that the true surface area is twice the estimated interface surface area.***

The method used for estimating the surface area is straightforward, and is rather like the method used for determining the volume. If we superimpose a straight line across a micrograph, it will cross any membranes that are present. The more membranes that there are, the greater will be the number of intersections and this can be quantified by counting. ***The surface density is therefore twice the number of intersections divided by the total line length***. If we reduce the line length to real cell dimensions by dividing by the magnification factor of the micrograph, the surface density will have the units $\mu m^2/\mu m^3$.

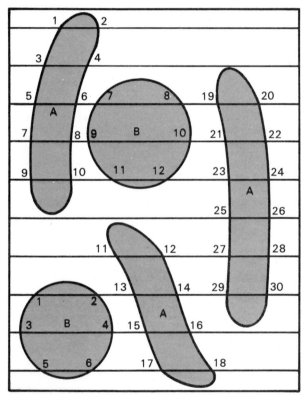

Figure 4.9
Estimating membrane surface area

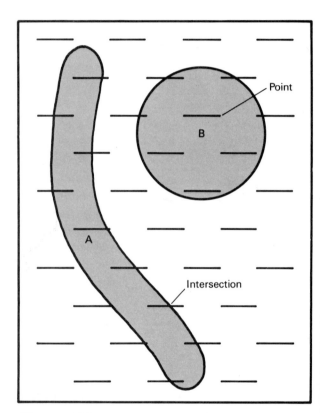

Figure 4.10
The multipurpose grid

The type of grid to use is shown in Figure 4.9. A single set of parallel lines are drawn, usually about 10 lines for a major grid and about 30 for a minor grid, and the total length of the lines within the grid is determined. In Figure 4.9 the grid consists of 10 parallel lines, each 8 cm long and 1 cm apart, giving a total line length of 80 cm. Two components, A and B, are shown. The total number of intersections with the boundary (and thus with the membrane) of A is 30, and of B is 12. The surface density estimate for A is therefore $2 \times 30/80 = 0.75$ cm^2/cm^3 and for B it is $2 \times 12/80 = 0.3$ cm^2/cm^3.

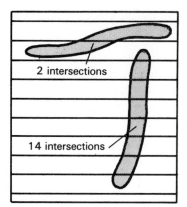

**Figure 4.11
Orientation of
components and lines**

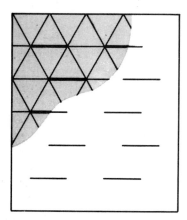

**Figure 4.12
Modified multipurpose
grid**

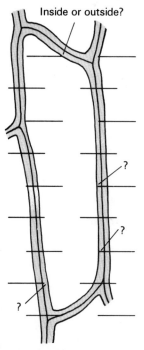

**Figure 4.13
Thick boundary (e.g. a
plant cell wall)**

There are difficulties with this method though, since the number of intersections depends on the orientation of the cell component with respect to the lines. If the component is lying parallel to the lines, then the number of intersections will be very small, whereas if it is lying at 90° to the lines, then the number of intersections will be high. See Figure 4.11.

The best way to overcome this is to count the intersections three times, turning the test grid by 120° each time.

We can combine the counting of both point and line intersects, and thus the calculation of volume fraction and surface area of components by using a *multipurpose grid*. This involves joining alternate pairs of dots of a point test grid using straight lines. The dots on adjacent rows are joined out of phase with the rows on either side so the end result is a test grid consisting of short lines. The ends of the lines are equivalent to the points of the point lattice and the lines form a test grid for intersect counts. See Figure 4.10. *If a line ends within a component, it scores one point; if a line crosses a membrane it scores one intersection*. The total number of points on the grid is twice the number of lines, and the total test line length is the number of test lines multiplied by the length of one of the lines.

This grid is composed of 35 lines, each of which is 1 cm long. This means that the number of points is $2 \times 35 = 70$ points, and the total line length is $35 \times 1 = 35$ cm. Two components, A and B, are shown. The number of points and intersections for each of the two components are:

	Points	Intersections
Component A	9	9
Component B	12	4

From these figures we can now calculate the total volume fraction (or percentage) and the surface density estimate.

The total volume fraction: A = 9/70 = 0.13 or 13%
B = 12/70 = 0.17 or 17%

The surface density estimate: A = $2 \times 9/35 = 0.51$ cm^2/cm^3
B = $2 \times 4/35 = 0.23$ cm^2/cm^3

This multipurpose grid can be improved by making the ends of the lines the same distance apart, both within and between the rows. This is achieved by basing the grid on equilateral triangles, as you can see in Figure 4.12.

Examine Acetate Sheet AS2, the minor squares grid; AS3, the major squares grid; AS4, the major points lattice; AS5, the minor points lattice; AS6, the multipurpose grid and AS7, the equidistant multipurpose grid.

These will be provided for you by your teacher. When you have selected the test grid that you want, it should be fixed firmly to the photograph or micrograph with paper clips so that it doesn't move about during the counting process. The lines are then scanned methodically one at a time, and the individual points or intersects counted and recorded. You might have a problem deciding whether a point lies inside or outside a component boundary. If a point lies directly on a boundary, count one in and one out alternately. If the boundary is quite thick, then you have to decide at the start if you are going to include it or not, so that you will count either to the inside or the outside edge of the boundary. See Figure 4.13.

Let us now look at an actual TEM micrograph to find out exactly how it is analysed.

Refer to Plate 2.10, on page 72, which is a TEM micrograph of a cell from a bean root tip, magnified 24 000 times. To make matters more straightforward, a line drawing was made of some of the components (nucleus, cytoplasm and vacuoles) by tracing from the micrograph. This is shown reduced in Figure 4.14(a), with the components labelled N, C and V.

An acetate sheet with a square lattice of points was then placed over the drawing (see Figure 4.14(b)). Points were counted over cytoplasm, nucleus and vacuoles and the results were as follows:

cytoplasm: 37

nucleus: 26

vacuoles: 8

total grid: 72

The first thing to do is calculate the area fractions. For each of these components it is:

cytoplasm: 37/72 = 0.514 or 51.4%

nucleus: 26/72 = 0.361 or 36.1%

vacuoles: 8/72 = 0.111 or 11.1%

Once these fractions are established, we can determine the volumes of the various components if we know the volume of one of them. Remember that areas, fractions and relative volumes are equivalent in ultra-thin sections. We can determine the actual volume of the nucleus by direct measurement. Since nuclei are almost spherical, a ruler was used to measure a few different diameters of the nucleus, and a mean value was calculated. This value in millimetres was then transposed into micrometres by measuring how long 1 μm was on the micrograph (note the scale given):

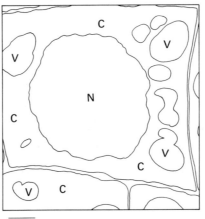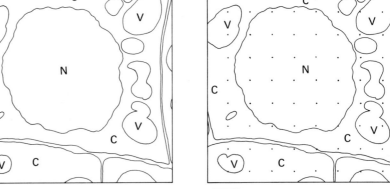

Figure 4.14
TEM micrograph of a bean
root tip cell

1 μm

a Outline drawing

1 μm

b With a point lattice

diameter of nucleus = 4.4 μm

thus radius R = 2.2 μm

so volume = $\frac{4}{3}\pi R^3$

= $\frac{4}{3} \times 3.14 \times (2.2)^3$

= 44.6 μm^3

We can express the volume fraction of the nucleus with respect to the cell, using the area fraction, since one is proportional to the other:

volume fraction of nucleus in cell = 0.361 μm^3/μm^3

The average cell volume is equal to the volume of the nucleus divided by the volume fraction of the nucleus in the cell:

$$= 44.6/0.361 = 123.5\,\mu\text{m}^3$$

The average cytoplasmic volume is the average cell volume minus the average volume of the nucleus:

$$123.5 - 44.6 = 78.9\,\mu\text{m}^3$$

Figure 4.15
Outline drawing of a bean root tip cell overlaid with a modified multipurpose grid

1 μm

The average vacuole volume per cell is the vacuole fraction multiplied by the average cell volume:

$$0.111 \times 123.5 = 13.7\,\mu\text{m}^3$$

The same micrograph was then overlaid with an equidistant interrupted line lattice based on a 1 cm isometric grid (Figure 4.15).

The results for points (ends of lines) are:

cytoplasm: 35

nucleus: 30

vacuoles: 11

total: 80

The use of line intersects is based on the knowledge that the plasma membrane lies between the cell wall and the cytoplasm, the nuclear envelope surrounds the nucleus and the tonoplast separates vacuoles from the cytoplasm.

The line intersect results are:

plasma membrane: 2

nuclear envelope: 10

tonoplast: 12

total line length: 40 cm = 40/1.25 (from scale) = 32 μm

The line intersect value for the plasma membrane is low because of the orientation of the lines and the cell walls. This can be overcome by rotating the grid relative to the micrograph, as mentioned previously, and taking three sets of readings.

From the points figures we can estimate the area fractions as a proportion or percentage of the relative volumes for the components, as with the point lattice.

$$\text{cytoplasm:} \quad 35/80 = 0.44 \text{ or } 44\%$$
$$\text{nucleus:} \quad 30/80 = 0.38 \text{ or } 38\%$$
$$\text{vacuoles:} \quad 11/80 = 0.14 \text{ or } 14\%$$

From the intersection figures we can estimate the surface densities since these are equal to twice the number of intersections divided by the total line length.

$$\text{plasma membrane:} \quad (2 \times 2)/32 = 0.125 \ \mu m^2/\mu m^3$$
$$\text{nuclear envelope:} \quad (2 \times 10)/32 = 0.625 \ \mu m^2/\mu m^3$$
$$\text{tonoplast:} \quad (2 \times 12)/32 = 0.75 \ \mu m^2/\mu m^3$$

We thus know the ratio of the nuclear surface to the other membranes. As with the volume of the nucleus, we can estimate the surface area of the nuclear envelope by using the measured radius (2.2 μm).

$$\text{surface area} = 4\pi r^2 = 4 \times 3.14 \times 2.2^2 = 60.8 \ \mu m^2$$

nuclear envelope area $\quad = 60.8 \ \mu m^2$

plasma membrane area $= \quad 2/10 \times 60.8 = 12.16 \ \mu m^2$

tonoplast area $\quad = 12/10 \times 60.8 = 72.98 \ \mu m^2$

Therefore, with very simple mathematical calculations, it is possible to estimate the actual surface areas of membranes and volumes of organelles in cells.

These stereological techniques can be used in a wide range of situations. Examples of these are given in Section 4.3.

4.2.3 Chromatography

Chromatography is an important technique in biology. It was developed in England in the 1940s by a team consisting of Consden, Gordon, Martin and Synge.

The stimulus to develop a new technique arose from the need to find out what amino acids were present in particular proteins. Paper chromatography resolved this problem by bringing about the speedy analysis of complex organic materials. It also made possible the separation and identification of extremely small quantities of substances of unknown mixtures – in contrast to the large quantities needed for classical techniques of purification and identification. The technique is so elegant and efficient that it has been applied to all sorts of compounds in almost every field of biology and chemistry.

Paper chromatography is basically a technique by which chemical substances can be separated and identified, using a moving solvent on sheets or strips of filter paper.

A drop or 'spot' of the solution containing the mixture of the substances to be separated is 'spotted' repeatedly about 3 cm away from one end of a piece of filter paper. The drop is then allowed to dry. The end of the paper nearest the spot is then placed in a suitable solvent, making sure that the spot is above the solvent level.

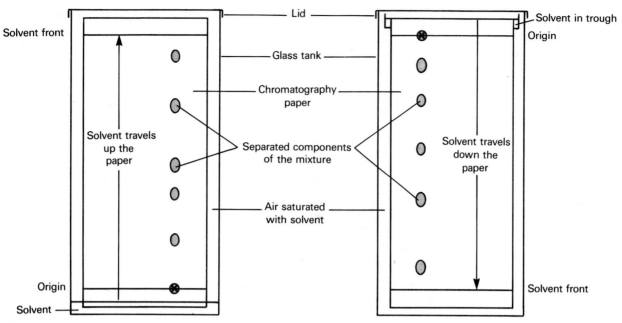

Figure 4.16
Ascending chromatography

Figure 4.17
Descending chromatography

In **ascending** chromatography the solvent is at the bottom of the vessel in which the paper is supported, and it rises up the paper by capillarity.

In **descending** chromatography the solvent is in a trough from which the paper is hung, and the solvent flows down the paper by a combination of capillarity and gravity.

In all cases the solvent flows along the paper, over and past the spotted mixture of substances. As it flows, it dissolves and carries along the substances from the spot, each substance moving at a different rate from the others.

The solvent flow is allowed to continue for a suitable length of time before the paper is removed from the tank and dried. This technique is known as **one-way** chromatography and the finished paper is known as a **one-way chromatogram**. If separation is difficult then the chromatogram is run twice, the second run being at right angles to the first run – this is called a **two-way chromatogram**.

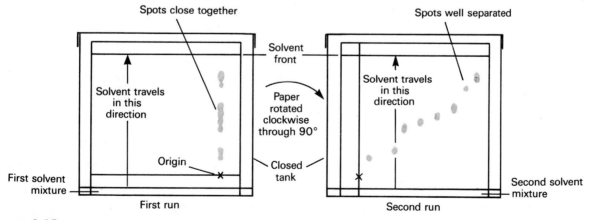

Figure 4.18
Two-way ascending chromatography

Why do the molecules of the mixture separate? Chromatography solvents are a mixture of water and an organic solvent such as ethanol or butanol. The cellulose of the paper has a stronger attraction for the water molecules than for the organic solvent so that a film of water accumulates on the surface of the paper (adsorption). This creates two phases:

(i)   a stationary phase involving the water, and

(ii)   a mobile phase involving the solvents.

Molecules of the mixture separate because they move at different speeds – their speed depending on how soluble they are in the two liquids. If a substance is very soluble in water and very insoluble in the solvent, then it will stay mainly in the stationary phase and hardly move at all. If, however, a

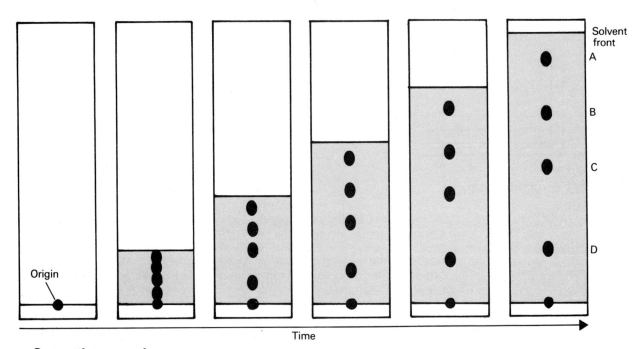

Time

a Stages in separation

Figure 4.19

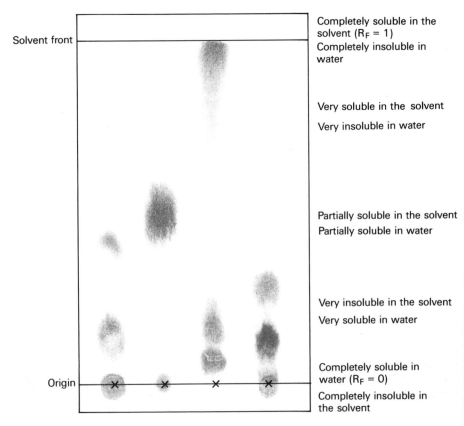

Solvent front — Completely soluble in the solvent ($R_F = 1$)
Completely insoluble in water

Very soluble in the solvent

Very insoluble in water

Partially soluble in the solvent
Partially soluble in water

Very insoluble in the solvent
Very soluble in water

Completely soluble in water ($R_F = 0$)

Origin — Completely insoluble in the solvent

b Typical chromatogram

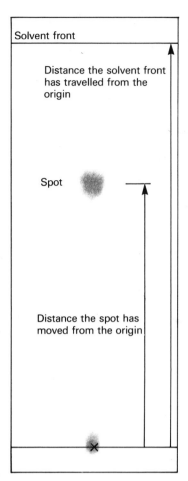

Solvent front

Distance the solvent front has travelled from the origin

Spot

Distance the spot has moved from the origin

Figure 4.20
Chromatogram showing distances the compound and solvent front have moved

substance is more soluble in the solvent than in water, it will spend more time in the mobile phase. Since the solvent moves up the paper, the molecules spending more time in the mobile phase move faster than those spending more time in the stationary phase, thus causing the separation.

R_F values

The furthest point reached by the advancing solvent on the paper is called the **solvent front**. This can be used as a reference point for describing the relative distances of travel of different substances in a chromatogram. The symbol used to designate this relative distance of travel is R_F (for relative front), and is defined as:

$$R_F = \frac{\text{the distance the compound has moved from the origin}}{\text{the distance the solvent front has moved}}$$

In order to determine the distance the solvent has moved from the origin it is always necessary to mark the solvent front with a pencil immediately the paper is removed from the tank, since the solvent evaporates rapidly.

The R_F value for the compound shown in Figure 4.20 is 70/107 = 0.65.

As the denominator is always larger than the numerator, an R_F value is a decimal number. Sometimes it is expressed as a percentage by multiplying by 100. The R_F is a useful figure because it is constant when all the experimental conditions are exactly reproduced, and can be used to characterise a particular compound. Of course, the R_F of a given compound will be different for different solvents. Other factors which affect the R_F value are the grades of paper used and the temperature at which the chromatogram is run.

1 Calculate the R_F values for the four compounds in Figure 4.19(a). Find out if the R_F values for each of the spots (A–D) are the same for each stage of the separation.

2 Repeat Question 1 for the four compounds in Figure 4.19(b). As the spots are not regular, the position of the spot is determined as the centre of the spot (see Figure 4.20).

Plate 4.1 is of a one-way ascending chromatogram of a leaf pigment extract. The following information is known: chlorophyll *a* is a blue-green colour whereas chlorophyll *b* is green; carotene is yellowish; phaeophytin yellow-grey and xanthophyll (luteol) yellow-brown.

The spots A to E were found to have the following colours:

> A: yellow
> B: yellow-grey
> C: yellow-brown
> D: blue-green
> E: green

3 Make out a table with the following column headings, and complete it.

Spot	Pigment	R_F value

PLATE 4.1

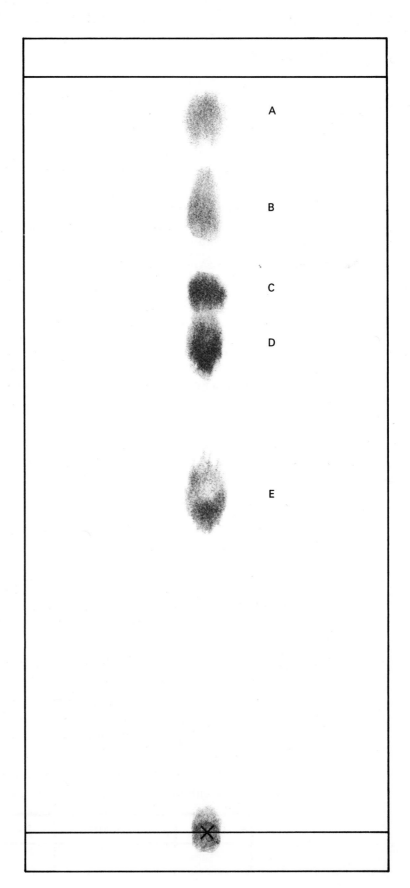

4.3 APPLICATIONS OF MEASURING

Now that you have achieved expertise in the skills of measuring and estimating try the following exercises.

4.3.1 Lengths of virus particles

Knowledge and skills required	Reference
Measuring	4.1.1
Recording	1.2.2
Calculating means	Vol. 1 p. 68
Calculating standard deviations	Vol. 1 p. 71

Plate 4.2 is a TEM of a virus called TRV. The magnification is ×190 000.

1 Measure and record the lengths of the virus particles to the nearest millimetre.

2 Convert your readings (in millimetres) into actual lengths (in micrometres).

3 Calculate the mean length and standard deviation of the sample.

4 Arrange the data so that you can record the information in a suitable graphical form.

5 Draw a graph of your data.

6 How would you describe the distribution you obtain?

7 Test your description by plotting the data on a suitable grid.

4.3.2 Banding in collagen fibres

Knowledge and skills required	Reference
Structure of collagen	2.2.3
Measuring	4.1.1
Determining means	Vol. 1 p. 68

In Section 2.2.3 you examined the structure of collagen and learned something of its function. Plate 4.3 is a TEM micrograph of non-sectioned collagen fibrils; shown on the plate is a $0.1\ \mu$m length.

As you can see, the collagen fibrils have a banding pattern which is similar in some ways to that found in skeletal and cardiac muscle – it indicates the arrangement of the constituent molecules, principally tropocollagen.

The banding consists of alternating dark and light bands in a very regular pattern. Three dark lines are just visible in the light bands.

1 Measure as accurately as you can the width of ten light bands to the nearest 0.1 mm.

2 Determine the mean width of the light band in nanometres.

3 Repeat for the dark bands.

PLATE 4.2

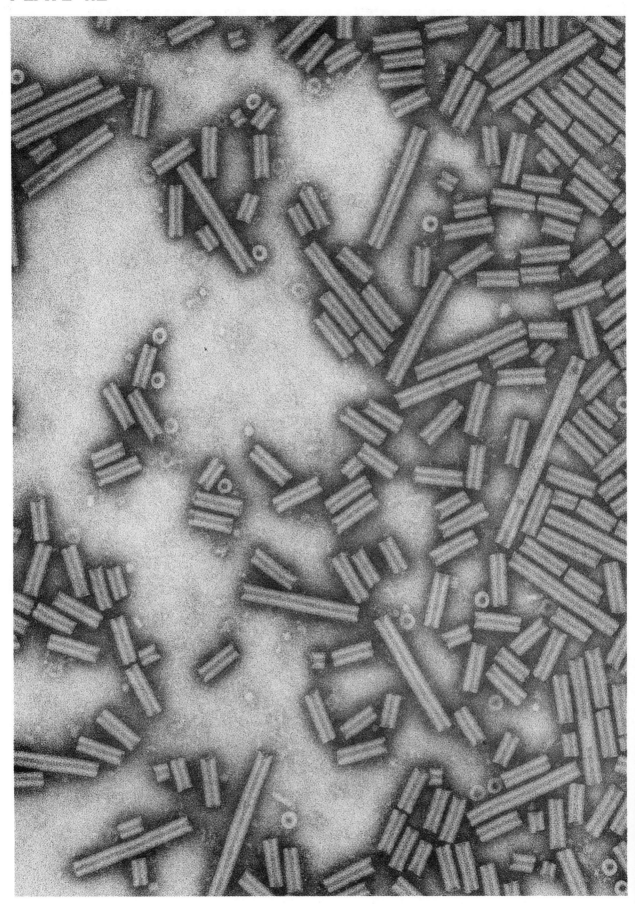

PLATE 4.3

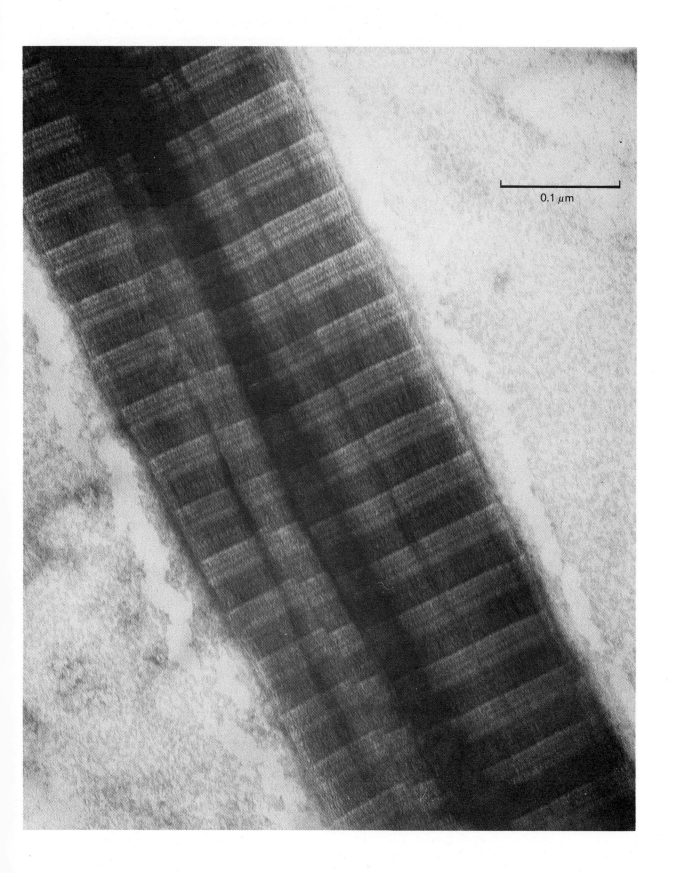

0.1 μm

4.3.3 Sizes of cells

Knowledge and skills required	Reference
Calculation of volumes of cuboids, spheres and ellipsoids	4.1.3
Calculation of surface areas of cuboids, spheres and ellipsoids	4.1.3
Calculation of means	Vol. 1 p. 68
Scales and magnification	1.1.9

Plate 4.4 is a micrograph of *Chlorella*, a single-celled alga which is roughly spherical in shape. The cell has been magnified 75 000 times.

1 What type of microscope was used to produce this micrograph?

2 Measure five diameters of the cell and calculate a mean diameter in millimetres.

3 Convert this mean figure in millimetres into a mean figure in suitable units.

4 Calculate the volume of the cell using the appropriate formula.

5 Calculate the surface area of the cell.

6 Repeat Questions 2 and 5 for the nucleus.

Plate 4.5 is a micrograph (×562) of plasmolysed onion epidermal cells. The cells can be considered to be cuboid in shape, while the cytoplasm can be taken to be ellipsoidal.

7 Measure a few lengths and a few widths of each of the three cells labelled 1, 2 and 3.

8 Calculate mean values of length and width for each of the three cells.

9 Calculate the volumes and surface areas of each of the three labelled cells.

10 Calculate the surface area to volume ratio for each of the three cells.

11 Repeat Questions 7 to 10 for the cytoplasm, taking the length as the *a*-axis and the width as the *b*-axis.

12 Arrange all the data in a suitable table.

13 The onion epidermal cells had been placed in a 0.9 M sugar solution. Why did the cytoplasm shrink?

14 What percentage of each of the three cells is occupied by the cytoplasm?

4.3.4 Micrometry of bean root cells

Knowledge and skills required	Reference
Techniques of micrometry	4.1.2
Calculation of mean	Vol. 1 p. 68
Calculation of variance	Vol. 1 p. 71
Significance tests	Vol. 1 p. 87

In the previous section (4.3.3), you were provided with highly magnified photographs to work from. Often these are not available and you would have to work directly with a microscope and slides. This section mimics situations such as that.

PLATE 4.4

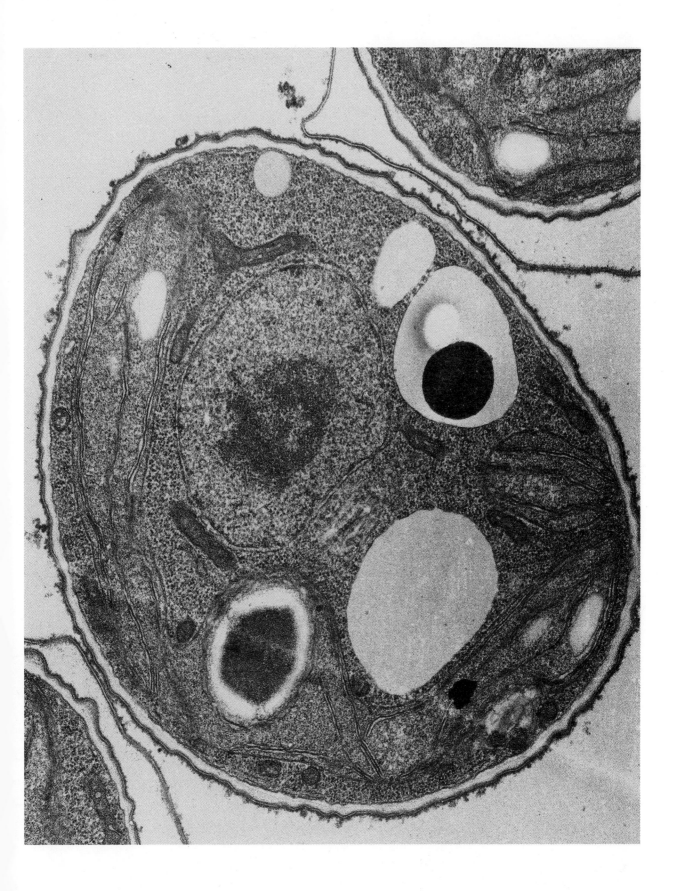

PLATE 4.5

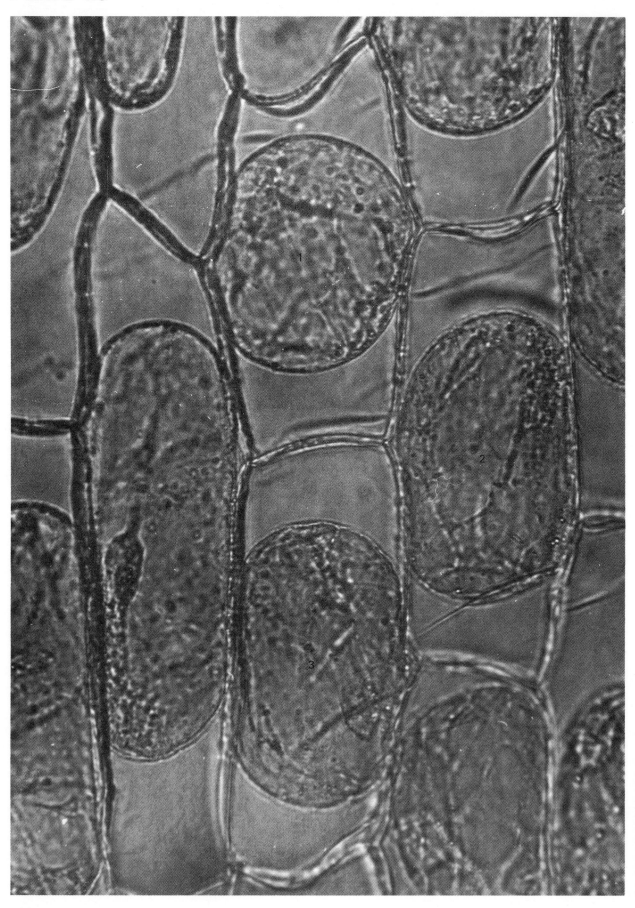

The drawings on Plate 4.6, overleaf are of longitudinal sections taken 1 mm, 2 mm and 5 mm from the apex of the bean (*Allium*) root tip. The drawings were taken from photographs all taken at the same magnification.

1 Make a labelled drawing of one cell from each photograph. Use the same scale for each drawing.

2 Make a list of the changes which you observe in the cells you have drawn as you move away from the tip.

You are provided with an acetate sheet (AS8) which has a scale (0–100) printed on it. This represents what you see when you look down a microscope fitted with an eyepiece graticule (see Sections 4.1.2 and 4.2.1). The scale on the acetate sheet can be used to measure the cells in the drawings (Plate 4.6) by superimposing the scale on top of the drawings.

3 For each of the drawings measure the lengths of five different cells.

4 Make up a table and record your results in it.

5 Calculate appropriate statistical parameters and record these in your table.

6 Outline a procedure which would have allowed you to select a sample of five cells at random.

7 List any practical problems which you encountered during the measuring and explain how you overcame them.

8 Are there any noticeable trends in the data? Give details.

9 Use an appropriate test of significance to analyse the data.

10 Repeat Questions 3, 8 and 9 for the widths of five cells from each drawing.

11 Repeat Questions 3, 8 and 9 for the diameters of the nuclei of five cells from the drawings.

12 Do the results obtained in Questions 8, 10 and 11 confirm your observations listed in Question 2?

13 All the measurements and calculations have used eyepiece graticule units and not SI units. Why is this justified?

14 At the magnification used for this exercise, one division of the eyepiece scale is equal to $10\,\mu$m. What is the mean length (in micrometres) of the cells measured in Question 3?

4.3.5 *Chlorella:* areas and volumes

Knowledge and skills required	Reference
Cell organelles	2.2.1
Stereological techniques	4.2.2
Estimating relative volumes	4.1.4
Estimating surface densities	4.1.4

In Section 4.3.3 we estimated the volumes of the cell and the cell nucleus by direct measurement of diameters. Let us now use the indirect methods employed by stereology.

PLATE 4.6

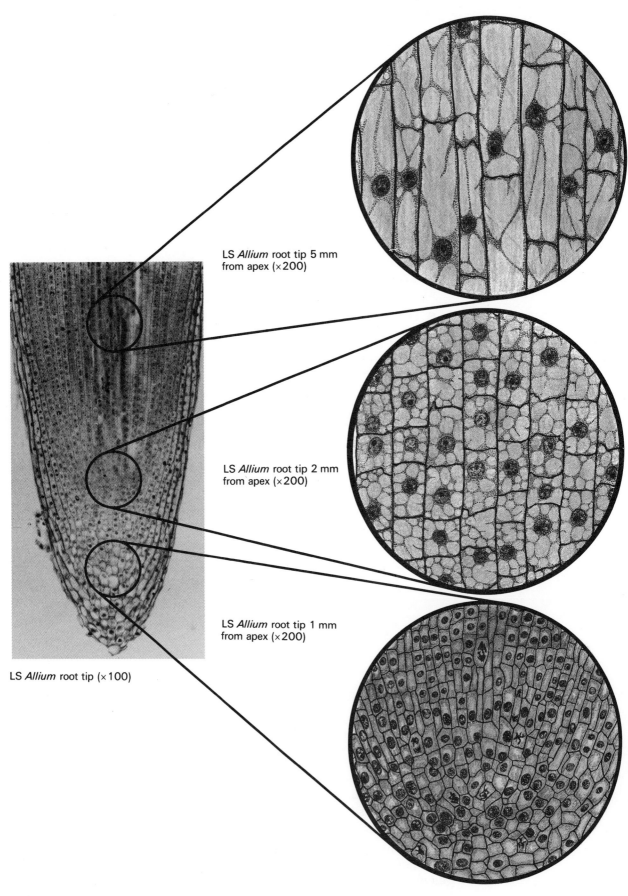

LS *Allium* root tip 5 mm from apex (×200)

LS *Allium* root tip 2 mm from apex (×200)

LS *Allium* root tip 1 mm from apex (×200)

LS *Allium* root tip (×100)

1 Place the equidistant multipurpose acetate grid, AS7 over Plate 4.4 (page 161), and secure it with paper clips so that it cannot move about. (You might find it easier to trace the outlines of the various structures from the micrograph and use this in place of the micrograph.)

2 As described on page 149, score the points and intersections for the single *Chlorella* cell. Score the points for the cytoplasm, nucleus, vacuoles and chloroplasts and the intersections for the plasma membrane, nuclear envelope, tonoplast and chloroplast envelope. Record the data in a suitable form.

3 Repeat Question 2 with the grid rotated 90° relative to the cell.

4 Calculate the mean points and intersections.

5 From the points scores, estimate the relative volumes of the
(i) cytoplasm
(ii) nucleus
(iii) vacuoles
(iv) chloroplasts.

6 From the intersections scores, estimate the surface densities of the
(i) plasma membrane
(ii) nuclear envelope
(iii) tonoplast
(iv) chloroplast envelope.

4.3.6 Plasmolysis in onion epidermal cells

Knowledge and skills required	Reference
Plasmolysis	Suitable text
Stereological techniques	4.1.4
Significance tests	Vol. 1 p. 87

In Section 4.3.3, you found out how to measure the size of onion cells. In this exercise we want to find the percentage volumes of cytoplasm in the cells at two different osmotic concentrations, this time using a dot matrix. The onion cells in Plate 4.7 (overleaf) were placed in an 0.8 M sugar solution, and those in Plate 4.8 were placed in a 1.0 M sugar solution.

1 Place the minor points lattice, AS5 over the micrograph Plate 4.7, and fix it with paper clips.

2 Count the points inside the cell walls of the cell and record the number.

3 Count the points in the cytoplasm and record the total.

4 Express the cytoplasm as a percentage of the total cell volume.

5 Repeat Questions 2, 3 and 4 for Plate 4.8.

6 Record your results in a suitable table.

7 Explain what has happened to the onion cells.

PLATE 4.7

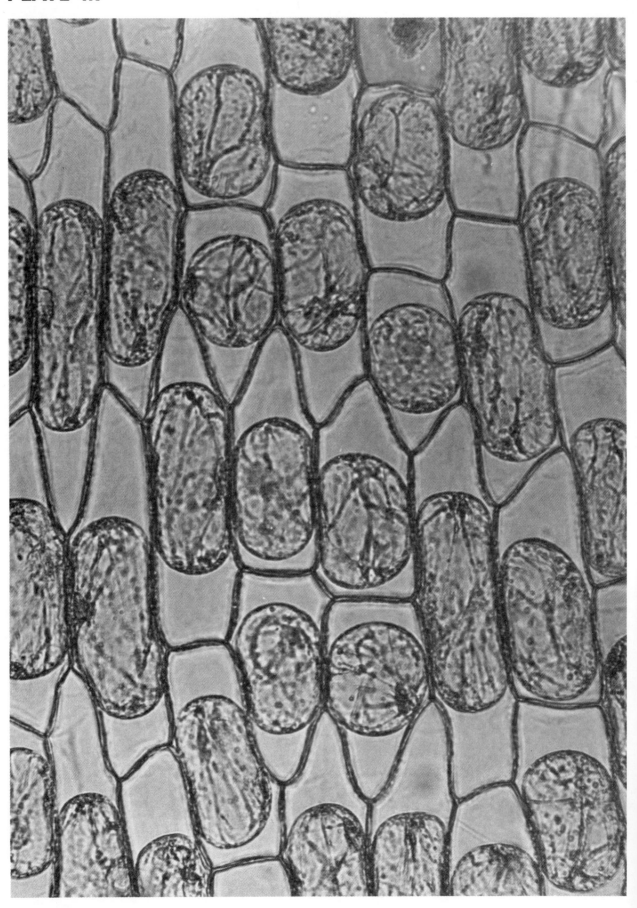

PLATE 4.8

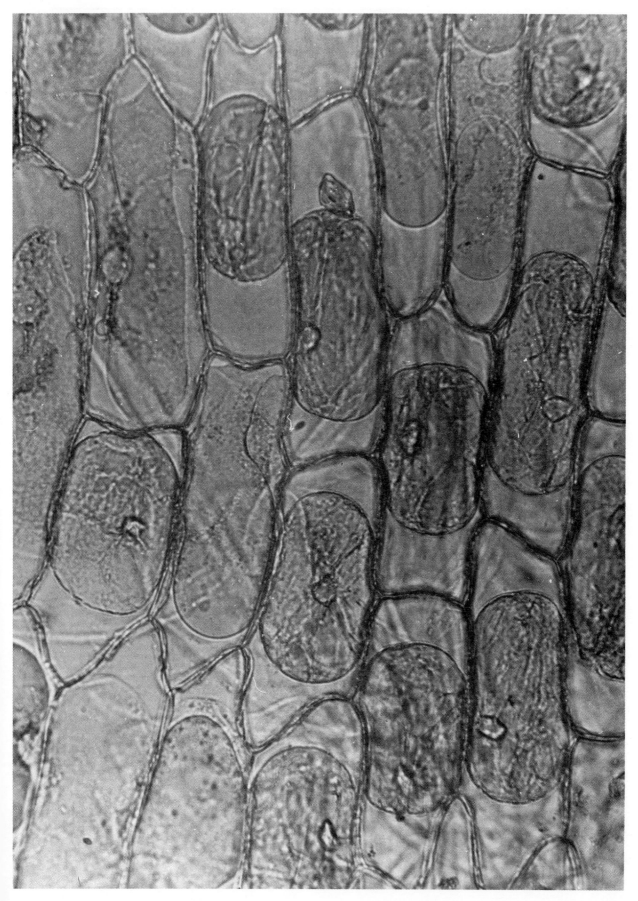

4.3.7 Capillaries in the lung

Knowledge and skills required	Reference
Structure of the lung	Suitable text
Structure of capillaries and red blood cells	Suitable text
Stereological techniques	4.1.4
Contingency tables	Vol. 1 p. 104
χ^2 significance test	Vol. 1 p. 96
Biological drawing	1.2.1

1 Examine Plates 4.9 and 4.10 which are of the blood–air barrier in the lung. Both plates show red blood cells inside capillaries.

2 Identify as many of the structures as possible and make a labelled and annotated biological drawing of Plate 4.9.

3 Examine Plate 4.10. Estimate, as accurately as you can, the distance from the surface of the red blood cell to the air in the alveolus.

4 Using appropriate stereological techniques, estimate the relative volumes of relevant structures.

5 Estimate the surface densities of relevant membranes using appropriate techniques.

6 Determine if there is any significant difference between the two plates.

4.3.8 Heart muscle

Knowledge and skills required	Reference
Cardiac muscle structure	2.2.3; suitable text
Magnification and scales	1.1.9
Stereological techniques	4.1.4

Plates 4.11 and 4.12 (pages 171–2) are TEM micrographs of cardiac muscle.

1 From the information given, work out the magnification in each case.

2 Using the appropriate technique, estimate the actual volume (in μm^3) of the nucleus in Plate 4.11, assuming that the nucleus is ellipsoidal in shape.

3 Estimate the relative volumes of the nucleus, muscle and mitochondria, in Plate 4.11.

4 Repeat Question 3 with Plate 4.12.

5 Using the appropriate significance test find out if there is any difference in the relative volumes of muscle and mitochondria between Plates 4.11 and 4.12.

4.3.9 The tongue

Knowledge and skills required	Reference
Magnification and scales	1.1.9
Mammalian histology	2.2.3
Stereological techniques	4.1.4

PLATE 4.9

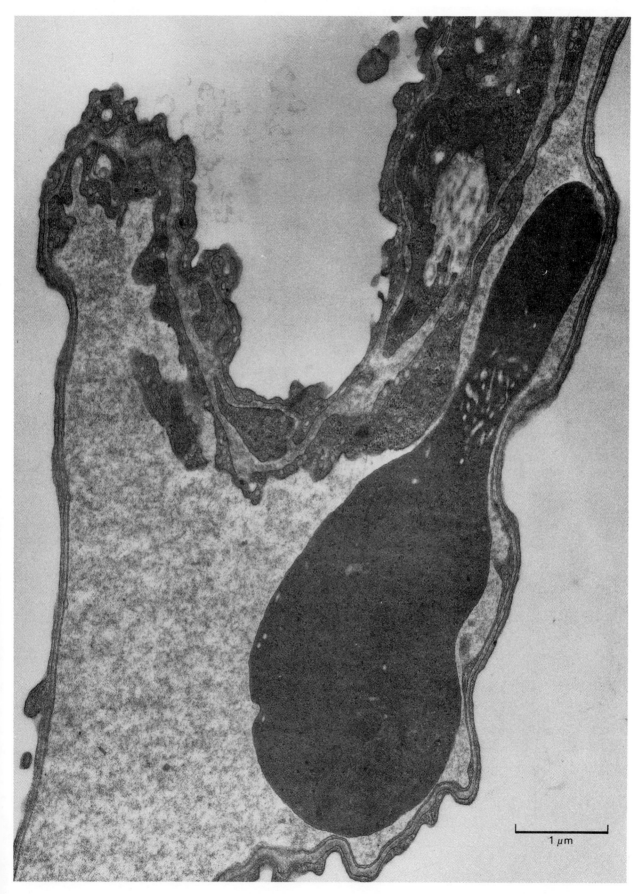

PLATE 4.10

PLATE 4.11

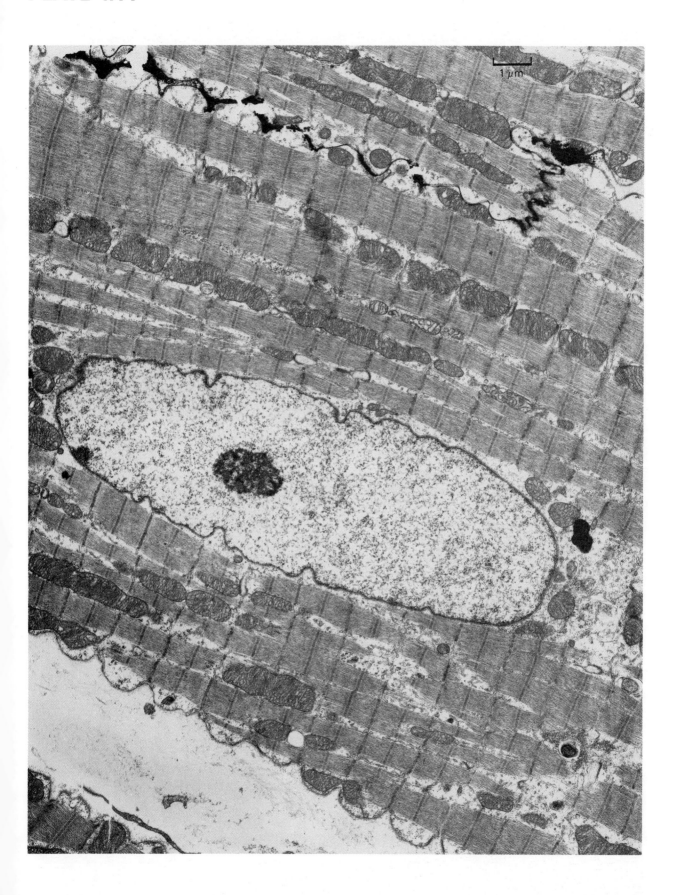

1 μm

PLATE 4.12

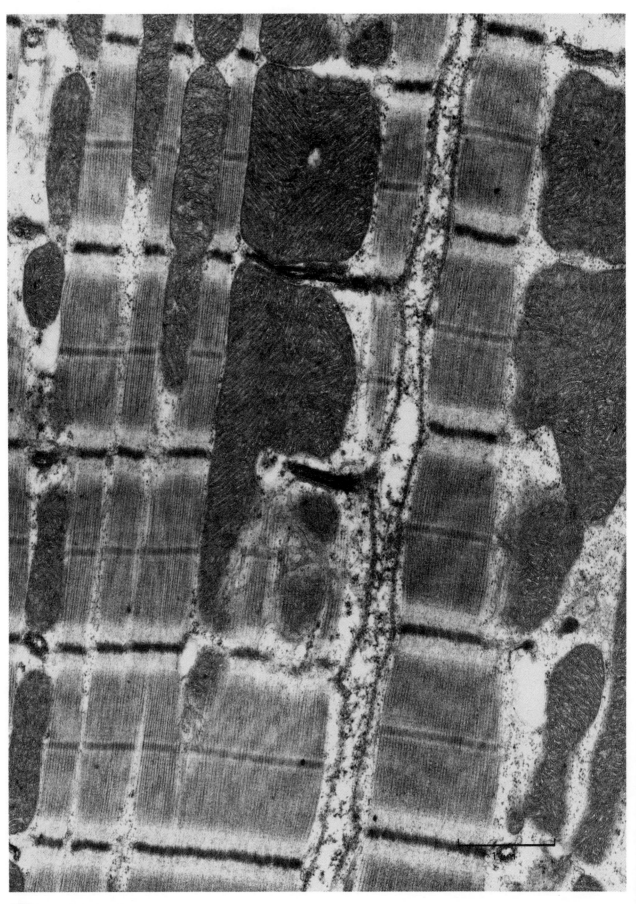

Plate 4.13 (overleaf) is a TEM micrograph of a section through the tongue.

1 What is the magnification of this micrograph?

2 Identify all the structures that are shown, and make a labelled and annotated biological drawing of the micrograph. The 'C' marked on the plate is collagen.

3 Estimate the relative volumes of the various structures you have identified.

5 Estimate the surface densities of the various membranes.

4.3.10 Other plates

Throughout this book we have used visual images of varying magnification or reduction. Many of these lend themselves to quantitative analyses of lengths, widths, areas, surface areas and volumes.

Examples of suitable plates and a suitable figure are as follows:

Plates: 2.1
2.2
2.3(a)
2.4(a) and (b)
2.8
2.9
2.10
2.11
2.13
2.14
2.18
2.21
2.24(b)
3.13
3.14

Figure: 1.33

PLATE 4.13

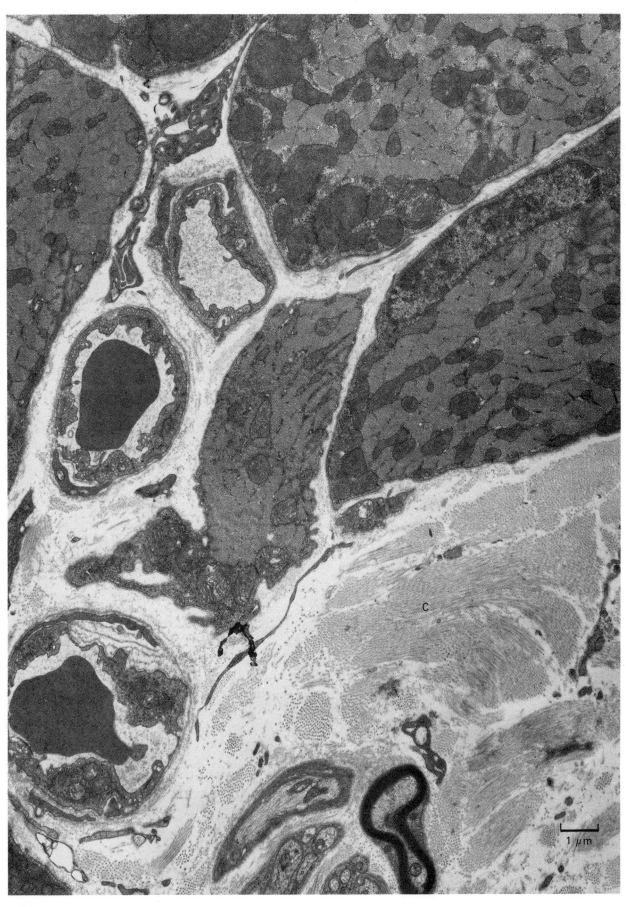

5 Observational Analysis

Occasionally analysis depends on nothing more than comparing certain items, and there is no requirement to count or measure. Careful observation requires attention to detail and looking for similarities as well as differences.

5.1 TECHNIQUES ALLIED TO PAPER CHROMATOGRAPHY

Various techniques which are essentially extensions of the principle of paper chromatography have been developed, and these have had a crucial role in various areas of biology. They are basically very simple techniques to use, but their potential for determining the presence of particular molecules from very tiny quantities, and for analysing biochemical pathways, has been phenomenal.

5.1.1 Thin-layer or gel chromatography

Paper chromatography is the separation of substances in a mixture on a thin layer of cellulose. It was found that thin layers of a wide variety of powdered inorganic materials such as silica gel and alumina, and organic substances such as starch and cellulose, could be used instead. Such gels had advantages over paper in that the separation time was much shorter and the spots were more clearly defined.

The adsorbent gel is deposited on a sheet of glass in a thin even layer and simply takes the place of paper. The glass plate is then placed in the solvent as for ascending paper chromatography.

R_F values are calculated in the same way as in paper chromatography.

5.1.2 Electrophoresis

If the substances in a mixture form ions, or if some of the substances ionise and others do not, separation of the different substances can be brought about by subjecting the mixture in solution to an electric field. This process is a form of electrolysis in which the movement of the products is stopped before they reach the electrodes. It can be carried out in much the same way as chromatography, using either paper or thin-layer gels. It is particularly useful in analysing proteins.

In an attempt to determine the type of fish that a fish-finger contained, the tissues of various species were subjected to electrophoretic separation on a gel.

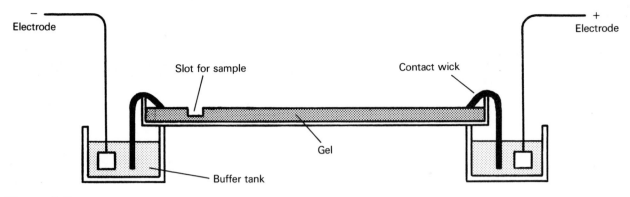

Figure 5.1
Electrophoresis

Figure 5.2 shows the electrophoretic separation of tissues on a gel for various species of fish.

1 – Nine-spined stickleback
2 – Atlantic salmon
3 – Brown trout
4 – Rainbow trout
5 – Arctic char
6 – Cod
7 – Whiting
8 – Rudd
9 – Roach
10 – Perch
11 – Eel
12 – Stone–loach
13 – Minnow

Figure 5.3 shows the same separation for a fish-finger.

1 What type of fish was the fish-finger made from?

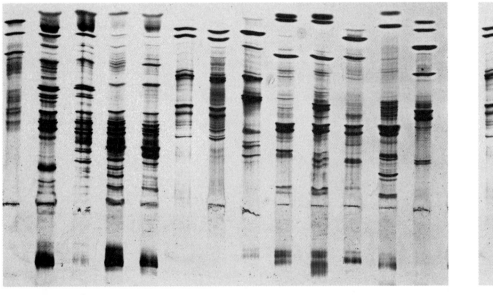

1   2   3   4   5   6   7   8   9   10   11   12   13

Figure 5.2
Electrophoretic gel of fish

Figure 5.3
Electrophoretic gel of
fish-finger

5.1.3 Autoradiography

This technique was developed to detect the presence of radioactive substances. Since radioactivity is a form of electromagnetic radiation, it will affect a photographic or X-ray plate in the same way as light or X-rays affect them.

Autoradiography is particularly useful when studying flat or two-dimensional structures like leaves or chromatograms, since the process depends on the material being in contact with an X-ray plate for a few days. The X-ray film is then developed and fixed as one would a photographic negative. Wherever the plate has turned black, this indicates the presence of a radioactive substance.

Using this technique in conjunction with paper or gel chromatography has been crucial in the working out of various biochemical pathways. An example of this was the analysis of the reactions taking place in the dark phase of photosynthesis by Melvin Calvin.

Calvin wanted to find out how the carbon dioxide taken in by plants was converted into organic substances like glucose. He reasoned that there would be a number of intermediate substances, and determining what these substances were, together with the order in which they were formed, would supply the answers.

$$Carbon\ dioxide \longrightarrow \underset{?}{A} \longrightarrow \underset{?}{B} \longrightarrow \underset{?}{C} \longrightarrow \underset{?}{D} \longrightarrow \underset{?}{E} \longrightarrow Glucose$$

This sequence of events occurs over time. The carbon dioxide would be converted into substance A, then A would be converted into B, B into C and so on. Calvin reasoned that if he supplied the plant material with carbon dioxide and then stopped the reaction after a short time, analysis should indicate what substance A was. Stopping the reaction after longer and longer intervals of time should reveal which substances were present at later stages.

Calvin used unicellular algae like *Chlorella* for his experiment. His 'lollipop' apparatus is shown in Figure 5.4 (overleaf). The alga was supplied with radioactive sodium hydrogen carbonate where the carbon was in the form ^{14}C instead of ^{12}C. At five-second intervals, a small volume of the algal culture was released through the tap at the bottom of the lollipop, to fall into a separate tube of hot alcohol.

1 Why did Calvin use unicellular algae for his experiment?

2 Why did he drop the samples into hot alcohol?

Each of the samples was then spotted on to chromatography paper and a two-way descending separation carried out. The resulting chromatograms were then placed in close contact with X-ray plates in the dark and autoradiograms produced.

3 Why was a two-way chromatographic technique used?

4 Why were autoradiograms produced?

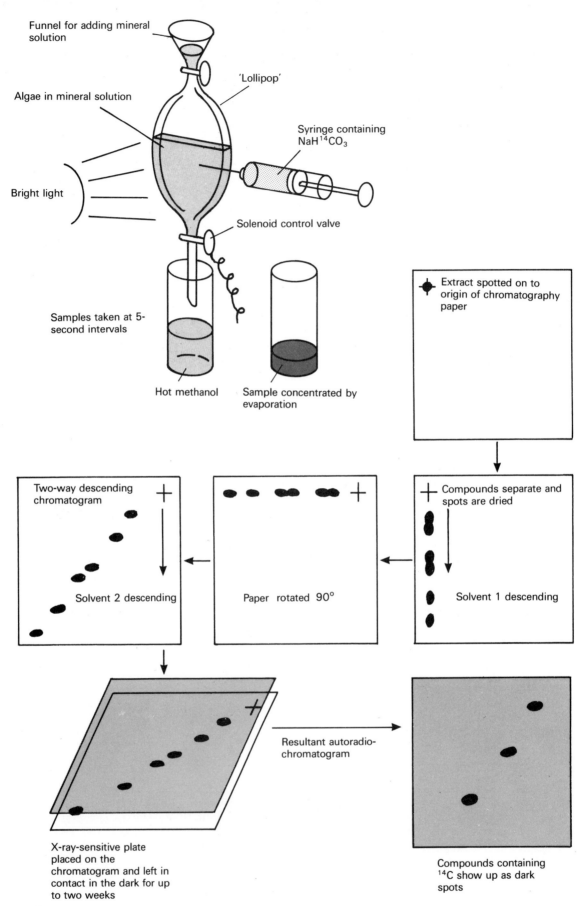

Funnel for adding mineral
solution

'Lollipop'

Algae in mineral solution

Syringe containing
NaH^{14}CO$_3$

Bright light

Solenoid control valve

Samples taken at 5-
second intervals

Hot methanol

Sample concentrated by
evaporation

Extract spotted on to
origin of chromatography
paper

Compounds separate and
spots are dried

Solvent 1 descending

Paper rotated 90°

Two-way descending
chromatogram

Solvent 2 descending

X-ray-sensitive plate
placed on the
chromatogram and left in
contact in the dark for up
to two weeks

Resultant autoradio-
chromatogram

Compounds containing
^{14}C show up as dark
spots

Figure 5.4
Calvin's experiment

Calvin's results were as follows:

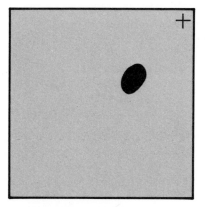

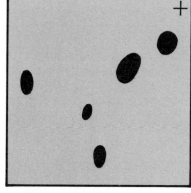

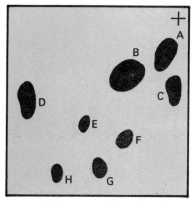

a After 5 seconds

b After 15 seconds

c After 60 seconds

Figure 5.5
Calvin's autoradiograms

5 What is the main difference between these three autoradiograms?

The substances represented by the black spots then had to be identified. Calvin achieved this by running two-way chromatograms (using the same solvents) of known substances which he thought might be likely candidates. He then compared their positions with those of the dark spots. Figure 5.6 shows a number of two-way chromatograms of different substances.

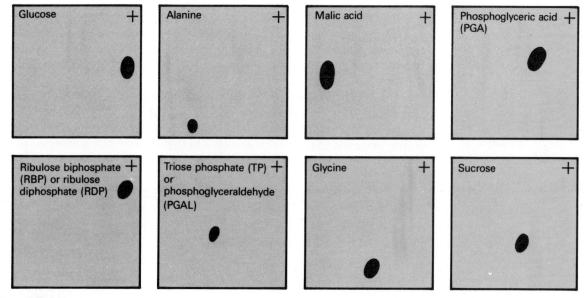

Figure 5.6
Two-way autoradio-chromatogram

6 Examine both Figures 5.5 and 5.6, and using the information given, work out what the substances A to H are. Arrange in tabular form.

7 From Figure 5.5(a), (b) and (c) determine the order in which the substances are formed.

From the information extracted from the autoradiograms, Calvin worked out that the ribulose diphosphate is regenerated so that the dark reactions are in the form of a cycle. This is now known as the Calvin cycle.

5.1.4 DNA sequencing

Knowledge and skills required	Reference
Structure and replication of DNA	Suitable text
Protein synthesis	Suitable text
Electrophoresis	5.1.2
Autoradiography	5.1.3

Fred Sanger was the first person to analyse the primary sequence of amino acids in a specific protein – insulin. After it became known that the sequence of amino acids is controlled by the sequence of bases in DNA, he was determined to develop a technique whereby the sequence of bases in DNA could be found.

Sanger's *dideoxy* technique was developed in 1975 and then modified in 1977. In it, single strands of DNA (which normally occurs as a double strand) are used as templates to produce complementary DNA of various lengths, containing a radioactive tracer. These single-stranded DNA fragments are then separated by gel electrophoresis. The resulting chromatogram is used to produce an autoradiogram which has a ladder-like banding pattern. From this the sequence of bases can be read directly, rather like the bar codes on goods. Recently the reading of such autoradiograms has been computerised and so can be carried out very rapidly. Since this whole process can be accomplished in a matter of weeks, it is faster to determine the amino acid sequence of a protein by this indirect route rather than by directly sequencing the protein, which is a very long, tedious process.

The whole process of sequencing will become clearer as you work through this section. Suppose that we need to determine the base sequence of a viral DNA because the virus is causing a serious disease. Knowledge of the structure of the viral DNA could assist in the development of a cure. The techniques used involve recombinant DNA technology, in which the viral DNA is extracted, purified and fragmented. Fragmentation is achieved by using specific *restriction enzymes*, which cut up sections of the DNA molecule. The fragments are then sorted and isolated. The portion of the DNA required is then attached to a carrier molecule and the hybrid DNA is introduced into a chosen cell in order to produce a suitable quantity for analysis.

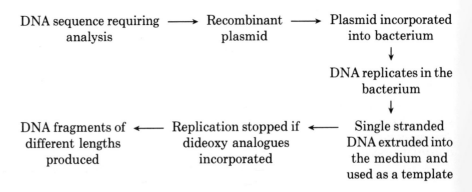

Figure 5.7
DNA sequencing

The genes are *cloned* (replicated to produce multiple copies) by using bacterial *plasmids* (small circles of DNA) which are smaller than the bacterial chromosome and can generally pass from one cell to another. By inserting the DNA into the plasmid ring, it can thus be introduced into the microbial host. Plasmids used in this way are known as *vectors* and certain viruses can perform this same role.

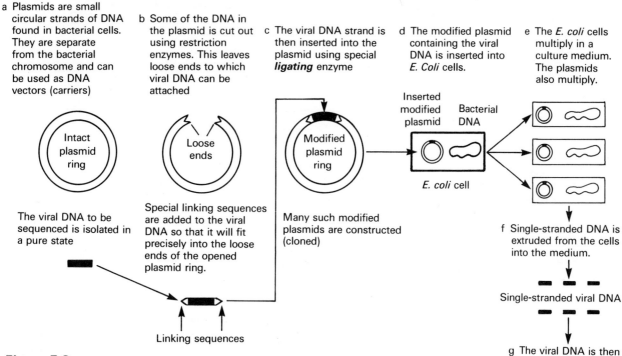

a Plasmids are small circular strands of DNA found in bacterial cells. They are separate from the bacterial chromosome and can be used as DNA vectors (carriers)

b Some of the DNA in the plasmid is cut out using restriction enzymes. This leaves loose ends to which viral DNA can be attached

c The viral DNA strand is then inserted into the plasmid using special **ligating** enzyme

d The modified plasmid containing the viral DNA is inserted into E. Coli cells.

e The E. coli cells multiply in a culture medium. The plasmids also multiply.

Intact plasmid ring

Loose ends

Modified plasmid ring

Inserted modified plasmid   Bacterial DNA

E. coli cell

The viral DNA to be sequenced is isolated in a pure state

Special linking sequences are added to the viral DNA so that it will fit precisely into the loose ends of the opened plasmid ring.

Many such modified plasmids are constructed (cloned)

f Single-stranded DNA is extruded from the cells into the medium.

Single-stranded viral DNA

Linking sequences

g The viral DNA is then sequenced

Figure 5.8
Obtaining single-stranded DNA

The cloned plasmids containing viral DNA are then inserted into living *Escherichia coli* cells. These plasmids are the replicated form, which is unusual because the plasmids extrude single-stranded DNA into the medium.

Using the extruded single-stranded DNA as a template, the sequencing method makes use of the ability of an enzyme called DNA polymerase to synthesise complementary copies.

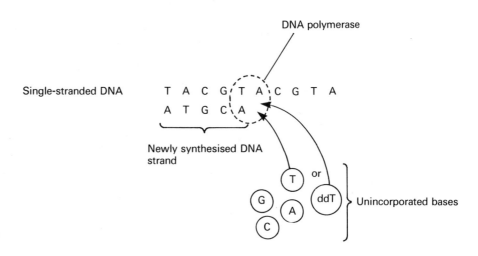

DNA polymerase

Single-stranded DNA   T A C G T A C G T A
                          A T G C A

Newly synthesised DNA strand

T or
G   ddT
   A
C

Unincorporated bases

Figure 5.9
Synthesis of DNA copies

Four reactions are prepared for each of the four bases (cytosine, guanine, thymine and adenine). The synthesis of DNA in each is stopped or 'terminated' by randomly incorporating one of four dideoxynucleotides, usually abbreviated ddC, ddG, ddT and ddA, corresponding to the four bases C, G, T and A. These dideoxynucleotides are the same as normal deoxynucleotides of DNA, except that they have one less oxygen atom. They are incorporated into the complementary DNA strands but their structure prevents any further nucleotides from being added. The result is that in each reaction there are lots of DNA strands of different lengths, all ending in the same dd base.

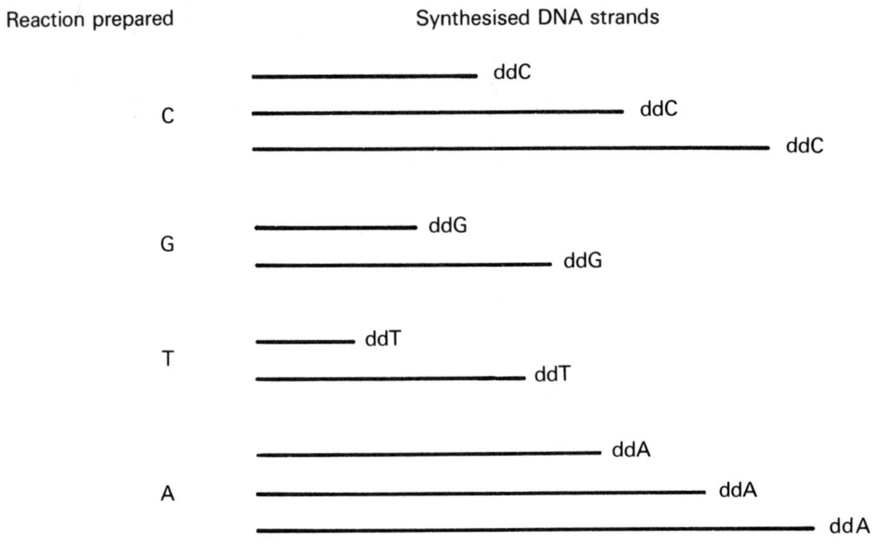

Reaction prepared | Synthesised DNA strands

Figure 5.10
Different DNA strand
lengths obtained for each
of the four reactions
corresponding to bases A,
C, G and T. Each strand
ends with its
dideoxynucleotide

For example, let us say that the original DNA strand was AACGGTACACG. The synthesised DNA strands will therefore have the transcribed code TTGCCATGTGC. If the chain terminator is dideoxy C, then this will become incorporated into the synthesised strand wherever the base C (cytosine) occurs:

$$\text{T T G C C A T G T G C}$$
$$* \quad * \qquad\qquad *$$

Thus, the following DNA strands will be synthesised:

T T G C C A T G T GddC
T T G CddC
T T GddC

In the same way, dideoxy G will become incorporated where the base G (guanine) occurs:

$$\text{T T G C C A T G T G C}$$
$$* \qquad\qquad * \quad *$$

so that the following DNA strands will be synthesised:

T T G C C A T G TddG
T T G C C A TddG
T TddG

1 Using the same base sequence, what strands will be synthesised that will terminate in ddT?

2 What strands will be made which end in ddA?

Introducing a radioactive nucleotide, for example cytosine nucleotide labelled with radioactive phosphorus (^{32}P) or radioactive sulphur (^{35}S), will make the newly synthesised DNAs of different lengths radioactive. These synthesised strands of DNA are then separated by electrophoresis on a polyacrylamide gel, in lanes corresponding to the bases C, G, T and A. The synthesised strands travel along the gel a distance which is inversely proportional to their molecular size and thus length: the shorter ones travelling further than the longer ones.

3 Which of the strands from the example will travel furthest along the gel?

4 Which strand will travel the shortest distance?

Since the strands are now labelled with a radioactive nucleotide they will show up as dark spots on an X-ray film, which is placed in contact with the gel and later developed as one would a photographic film. The resulting autoradiogram would look something like this for the DNA strand in question.

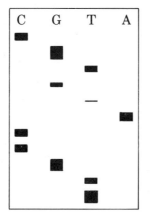

Reading up from the bottom of the autoradiogram this will be the sequence of bases

5 What would be the base sequence in the following autoradiogram?

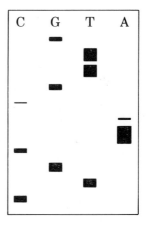

Plate 5.1(a) is an autoradiogram of bovine enterovirus. Plate 5.1(b) shows a small part of it enlarged.

6 See if you can find the short enlarged sequence (b) in (a). (The small dot artefact should help you.)

7 What is the sequence of bases in the boxed portion of the central lane of (b)?

8 Write out the DNA sequence obtained in Question 7 in triplets starting from the left, and convert it into the complementary RNA sequence.

Three bases taken together (known as a triplet codon) code for one amino acid, specific triplet codons coding for specific amino acids. The complete code is given in Figure 5.11.

PLATE 5.1

a

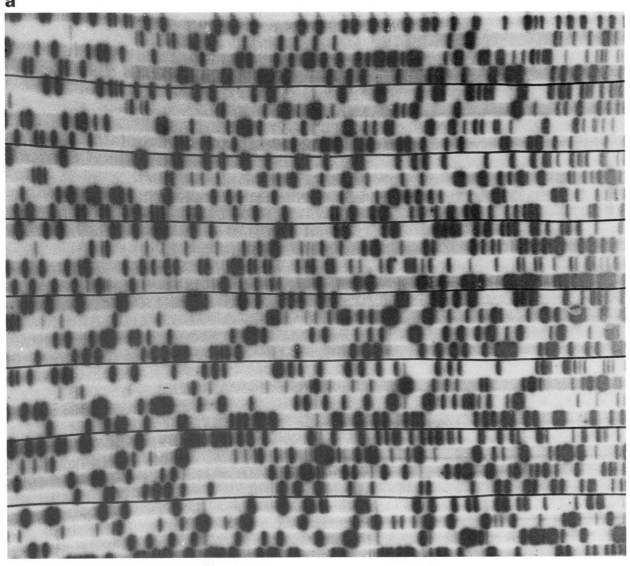

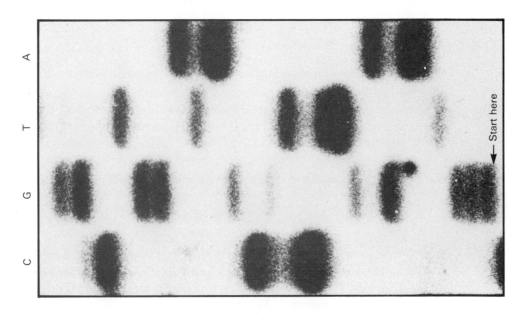

Triplet codons for amino acids

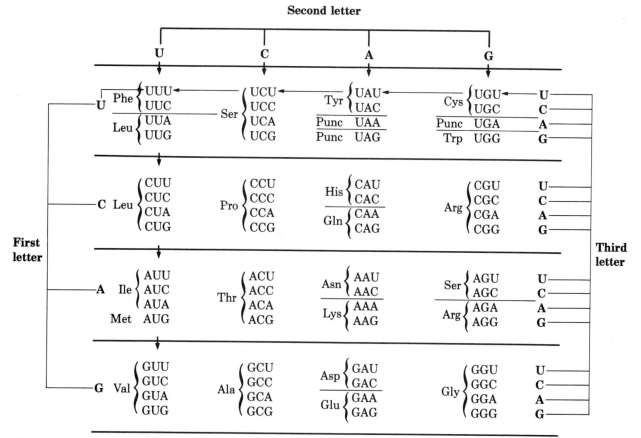

The three codes which do not code for amino acids act like punctuation marks. UAA and UAG seem to work like genetic full stops.

Phe – phenylanine	Thr – threonine	Asp – aspartic acid
Leu – leucine	Ala – alanine	Glu – glutamic acid
Ile – isoleucine	Tyr – tyrosine	Cys – cysteine
Met – methioine	His – histidine	Trp – tryptophane
Val – valine	Gln – glutamine	Arg – arginine
Ser – serine	Asn – asparagine	Gly – glycine
Pro – proline	Lys – lysine	

Figure 5.11
The genetic code

9 Using this code, translate the RNA strand you obtained in the previous question into the amino acid sequence of the polypeptide that was originally coded for by the sequence of the viral DNA.

A technique known as DNA profiling was developed by Alec Jeffreys in 1985. It produces a banded autoradiogram when human DNA is cut up with particular restriction enzymes and separated on an electrophoretic gel. This technique provides invaluable information in forensic cases, where human DNA obtained from body fluids such as blood or semen can be matched with suspects' DNA. Only minute amounts are required for the analysis; cheek cells in saliva can even be used, and if a minimum of 14 'bars' on the genetic 'bar code' can be matched then there is no doubt that the DNA from the two samples belong to the same person. Since DNA profiles are unique to each individual, the technique has become known as ***DNA fingerprinting***. This technique can also be used to determine the parentage of a child, whether in cases of disputed parentage or in immigration issues.

5.2 CLASSIFICATION AND KEYS

Arranging, grouping or classifying is a human trait which begins in young children. They recognise and organise the messages from their sense organs into categories of similar impressions, for example soft, warm, tall, red, etc. Any new experience is compared to the categories already established, reinforcing them or laying the bounds of a new division. Eventually a mental framework or definition is established in the brain cells and for ease of reference each category is given a name, for example bicycle, motorbike, car, van, lorry, train.

Living things are very difficult to classify, both because they change morphologically during their lifetimes, and because there are in excess of one million known species.

Informal classifications based on common names can be inaccurate and confusing. For example, the word **fish** appears in crayfish, shellfish, and starfish, all of which live in water, but which have few common structural features. There are several different ways of grouping living organisms. The two methods outlined below each have a different purpose.

The most complex classification system is termed a **formal** or **natural** classification. This is how biologists attempt to detail information about all the known types of living thing, the **species**. Those species which have more characteristic features in common than they have differences are believed to be related and thus belong to a larger grouping, the **genus**. Where several genera have a majority of features in common they in turn are grouped together at the **family** level. Similar grouping on this hierarchical basis leads to larger and larger groups which emphasise the common features of more and more species, through the family, to the **order**, **class**, **phylum** and **kingdom**. The formal classification is an internationally agreed system and once the features typical of each level or group are known, they can be applied to all the other members of the group, allowing predictions to be made.

Informal or **artificial** systems are used mainly for the purpose of identifying species from among a precisely defined community of species. This community may be comprised of the organisms found in a freshwater pond or in a stone wall, the flying insects, or the bacteria known to cause human illness.

Figure 5.12
Leaves found in a length
of hedgerow

Horse chestnut   Ash   Oak   Sycamore

Guelder rose   Hazel   Privet   Hawthorn   Beech

The simplest method of devising an artificial classification is to divide the defined group of organisms into two subgroups or subsets on the basis of one easily observed character. Exactly which character is used depends on the person making up the classification – two students working on the same group of organisms may decide on very different, but equally correct, criteria.

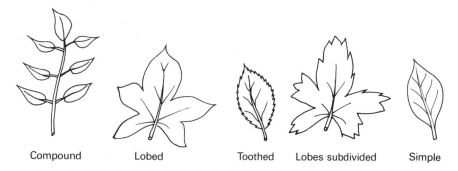

Figure 5.13
Characteristics used to describe leaves

Compound   Lobed   Toothed   Lobes subdivided   Simple

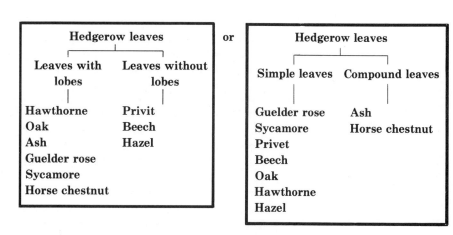

Figure 5.14
Alternative divisions of leaves into subsets

The procedure is repeated with each of the subsets, using a different criterion in each case, until all the types of organism have been separated into their own subsets. The classification can now be written down in the form of a branching tree, with each trunk dividing into two. This is known as a **dichotomous** system and is often used for identification purposes.

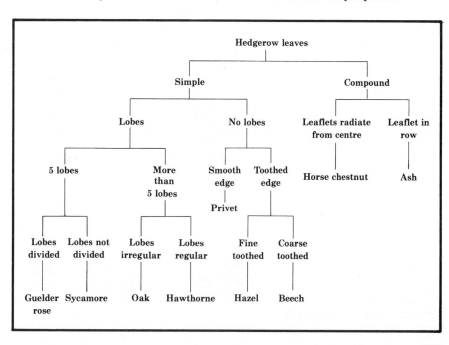

Figure 5.15
Dichotomous classification of leaves

More often however, the classification is reorganised using a series of key questions arranged in pairs, with only one question in each pair having a positive answer for any particular organism. Set out in this fashion the classification is termed an ***identification key***.

1.	Leaves simple (not divided into leaflets)	If YES go to	2
	Leaves compound (divided into leaflets)		8
2.	Leaves divided into lobes		3
	Leaves not divided into lobes		6
3.	Leaves with five lobes		4
	Leaves with more than five lobes		5
4.	Lobes subdivided into smaller lobes	Guelder rose	
	Lobes not subdivided	Sycamore	
5.	Lobes irregular	Oak	
	Lobes regular	Hawthorne	
6.	Edge of leaf smooth	Privet	
	Edge of leaf not smooth (toothed)		7
7.	Edge of leaf coarsely toothed	Beech	
	Edge of leaf finely toothed	Hazel	
8.	Leaflets radiating from a central point	Horse chestnut	
	Leaflets arranged in rows	Ash	

**Figure 5.16
Identification key for
leaves**

1 Plate 5.2 shows eight different types of arthropod. Using the procedure outlined above which includes
(i) tabulation of visible characters,
(ii) construction of the classification, and
(iii) rearrangement as a key,
devise an identification key for the group.

PLATE 5.2

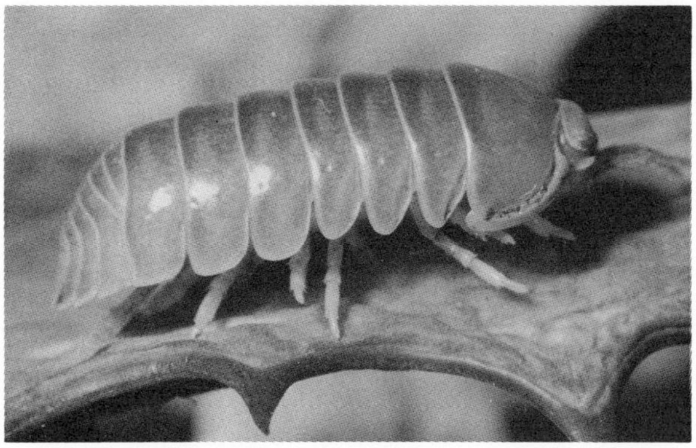

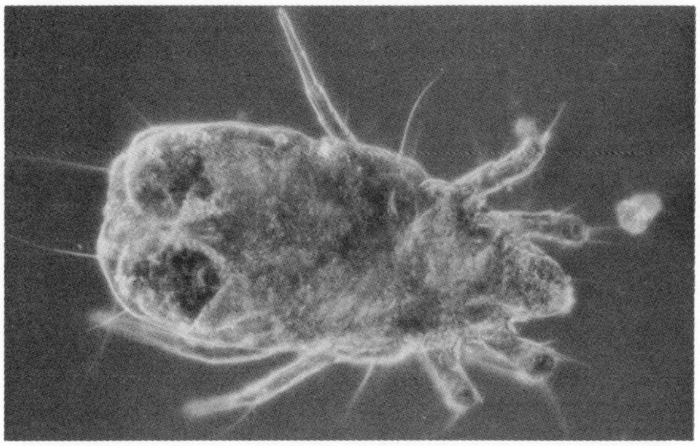

Index

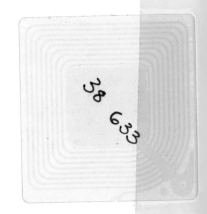